Listen.
My Child Has a Lot of Living to Do

The partnership between parents and professionals in
caring for children with life-threatening conditions

Edited by

J. D. Baum,
Sister Frances Dominica, and
Robert N. Woodward

Published in association with the
Institute of Child Health, Bristol

OXFORD UNIVERSITY PRESS
Oxford   New York   Tokyo
1990

Oxford University Press, Walton Street, Oxford OX2 6DP

Oxford   New York   Toronto
Delhi   Bombay   Calcutta   Madras   Karachi
Petaling Jaya   Singapore   Hong Kong   Tokyo
Nairobi   Dar es Salaam   Cape Town
Melbourne   Auckland

and associated companies in
Berlin   Ibadan

Oxford is a trade mark of Oxford University Press

Published in the United States
by Oxford University Press, New York

British Library Cataloguing in Publication Data

Listen. My child has a lot of living to do: the partnership
between parents and professionals in caring for children
with life-threatening conditions.
1. Terminally ill children. Care
I. Baum, J. D. (John David)   II. Francis Dominica, Sister
III. Woodward, Robert N.   IV. Series
362.1750834
ISBN 0–19–261898–9
ISBN 0–19–261961–6 (pbk)

Library of Congress Cataloging in Publication Data
(Data available)

ISBN 0 19 261898 9
ISBN 0 19 261961 6 (pbk)

Set by BP Integraphics Ltd, Bath
Printed and bound in Great Britain
by Biddles Ltd, Guildford and King's Lynn

Foreword

I am very glad to have the opportunity to welcome this anthology of the experiences of children with life-threatening illness and of their families.

The book reveals with gentle discernment the personal and individual needs of afflicted children and families. But it also confirms that the burdens imposed by unrelenting illness, the threat of premature death, the limitations of medical treatment, and the need for spiritual, emotional, and physical support are common to them all.

Often those needs have been associated with children's hospices. A hospice is commonly and misleadingly thought of as a place to die; in this sense the word has lost its original meaning as a place of rest and sustenance on a journey. Thus in the circumstances of life-threatening illness in childhood 'hospice' describes the diverse forms of support and respite available to the family and the ill child. This richer usage is the underlying theme of the book.

An essential step towards better understanding is to learn from the children and their families, and those who have close involvement with them, something of their experiences and views. That is the purpose of the book and it is a purpose I strongly support.

Sir Donald Acheson
Chief Medical Officer
1990

Preface

There is infinite diversity among children with life-threatening and terminal illnesses: they do not represent a neat homogeneous category of patients. The nature of their illnesses varies widely as does the progression of their disease, the location of their home in relation to their medical services, family circumstances, finance, aspirations, and attitudes to life and death. However, for all these differences, we believe these children have a number of features in common: the inexorable progression of their disease process; the imponderable emotional burdens of premature mortality; and the balance between the child's need for technical and scientific medical care on the one hand and his family's need for emotional and spiritual support on the other.

Over the last decade, a realization has dawned, predominantly among voluntary groups associated with the experts in this field—namely, the children and families whose lives have been affected by progressive and terminal illness—that between the bricks and mortar of hospitals and the established community services, large gaps exist in the provisions of care for which these families feel a vital need. Once identified, these needs were seen to be urgent and pressing and yet largely without models on which to build.

In this book we have brought together descriptions of some of the initiatives that have been taken to redress these deficiencies. The spectrum of the problems and their possible solutions is such that no single text can possibly hope to be comprehensive. Thus, while recognizing that there are particular problems associated with individual progressive disorders, we have selected examples to illustrate general points rather than try to cover each specific organ failure and disease.

This book is not intended as a criticism of medical and social services in this country. It is meant rather to represent something of the families' tragic experiences and their expert views on how existing services might in the future be improved to provide a comprehensive, flexible, and compassionate national network of care for the families and their children afflicted by life-threatening diseases.

J. D. B.

Institute of Child Health, Bristol
1990

Contents

Editors

J. D. Baum
Professor of Child Health, University of Bristol
Director, Institute of Child Health, Bristol

Sister Frances Dominica
Director, Helen House, Oxford

Robert N. Woodward
Chairman, CLIC, Director, CLIC UK, Bristol

Managing Editors

Jill Pomerance and Jo Hearn
Research Associates, ACT, Institute of Child Health, Bristol

Contributors

David Baum MA, M.Sc., MD, FRCP
Professor of Child Health, University of Bristol, Director of the Institute of Child Health, Bristol.

Bernadette Cleary
Founder of Rainbow Trust, Great Bookham, Surrey.

Sally Curnick SRN, RSCN, NDN, Dip.Hum.Psy.
CLIC Domiciliary Nurse, Bristol.

Sally Day RGN, RSCN
Director, Acorns Children's Hospice, Birmingham.

Ann Dent SRN, RCNT
Cancer Relief Macmillan Fund Nurse Consultant for children with life-threatening diseases, Macmillan Education Centre, Dorothy House, Bath.

Sister Frances Dominica RGN, RSCN, FRCN
Founder and Director, Helen House, Oxford.

Gillian C. Forrest MB, BS, F.R.C.Psych., DCH, D.Obst., RCOG
Consultant Child Psychiatrist, The Park Hospital for Children, Oxford.

Ann Goldman MA, MB, MRCP
Director of Symptom Care Team, Department of Haematology and Oncology, Hospital for Sick Children, Great Ormond Street, London.

Owen Hagen BA, CQSW
Senior Social Worker, Royal Liverpool Children's Hospital, Alder Hey, Liverpool.

John Halliday MA, MB, B.Chir., MRCP
General Practitioner, South Brent, Devon

Jo Hearn
Research Worker ACT, Institute of Child Health, Bristol.

Eve Herd BA
Deputy Head, Helen House, Oxford.

Lenore Hill RGN, RSCN
Head Nurse, Martin House, Boston Spa, Wetherby, W. Yorks.

Peter Jeffrey
Head Teacher, Brookfield House School, Woodford Green, Essex.

Hyam Joffe MD, FRCP, M.Med.(Paed.), FACC
Senior Clinical Lecturer, Consultant Paediatric Cardiologist,
Royal Hospital for Sick Children, Bristol.

Christine Lavery
Honorary Director, MPS Society for Mucopolysaccharide
Diseases, Development Officer Rare Handicaps Groups, Contact a Family, Secretary Genetic Interest Group (GIG).

Jacqueline Mok MD, MRCP, DCH
Consultant Paediatrician, Department of Community Health,
Edinburgh, and Department of Infectious Diseases, City
Hospital, Edinburgh.

Tina Neale BA, B.Phil.
Project Leader, Barnardos Cystic Fibrosis Project, West
Midlands.

Alan Stein MB, B.Ch., M.R.C. Psych.
Wellcome Trust Lecturer, Department of Psychiatry, University
of Oxford.

Jillian Tallon
National Secretary, The Compassionate Friends.

Rosemary Thornes BA, M.Sc.
Project Officer, Caring for Children in the Health Services Committee, Bristol.

Robert L. Woodward LL.D
Founder and Chairman of CLIC (Cancer and Leukaemia in
Childhood Trust).

Helen Woolley BA, Dip.Soc.Admin., PSW
Research Associate, Department of Psychiatry, University of
Oxford.

Helen Vegoda CQSW, Cert.Ed., Dip.Soc.
Paediatric Cardiology Counsellor, Royal Hospital for Sick
Children, Bristol.

Beth's picture

JO HEARN

This picture was drawn by my daughter Beth when she was seven and a half, just a few weeks before she died at Helen House in Oxford.

Beth had been ill for much of the preceding four and a half years and knew, in a child's way, that she did not have long to live.

It was so very important to us that every tiny second we had left together should be as good, as near to perfect, as it could possibly be. At home we played beautiful music and the house was full of flowers; there were little children to tea, and close friends and family sharing some of our days . . .

But at home, too, in those last few weeks, despite the efforts of the local medical and nursing teams there were awful crises; a vertebra collapsed and paralysis from the waist down resulted; there was pain, appalling pain, constipation, vomiting, oral thrush, urethral catheters, diarrhoea, more pain, exhaustion all round, fear . . .

At Helen House the fear diminished. Symptoms were dealt with as they began to manifest themselves, or were pre-empted by a confident and positive approach to symptom control, involving combinations of drugs which worked.

Our trust was slowly and patiently won as the house team became our friends. We were 'held', cossetted, loved. We were cushioned against impossible reality, and as the fear subsided and we felt safe again, it was possible and permissible to give our energy to Beth.

One morning, at Helen House, I was sitting next to Beth, looking at her frail body as she dozed, and I wept; she opened her eyes and said: 'Why are you crying Mummy?' I said 'I'm crying because you are not well; I'm crying because I so wish that I could wave a wand and make you well, so that you could jump out of bed and run and dance down the corridor like any other little girl ... just like you used to do. I would give anything in the world to make that happen.'

And Beth took my hand and she said:

'It's *all right* Mummy, it's *alright*'

Beth died in my arms and my husband's, in a huge double bed that had been specially made up for us on the floor of Beth's room at Helen House; our other children David, Beth's twin, and little Jennie were safely asleep in the adjoining room.

There was peace.

Part I

Introduction

1 Hospices: a philosophy of care

FRANCES DOMINICA

In another culture, or in a different era, hospice care for children might well be considered redundant. In many ways it is a poor replacement for the extended family. Although families in previous generations frequently faced wasting and death of children, and even today those in Third World countries continue to do so, the existence of the close-knit extended family and the greater involvement of the local community provided a kind of support which is generally lacking in contemporary British society. The nuclear family, not infrequently a single-parent family, can suffer terrible loneliness and isolation. Its members may not have sufficient resources to meet each other's physical, emotional, or spiritual needs during the child's illness, death, and aftermath. Consequently, there is a role for others, outside the immediate family, to be alongside, offering friendship, support, and practical help, however protracted, throughout the child's illness and during both the terminal phase and the months and years of bereavement that follow.

Professional caregivers always run the risk of taking charge of the situation, to the diminishment and often the frustration of the family concerned. It behoves us to recognize from the start the expertise of the family in caring for their own sick child, and, in many cases, the expertise of the child him or herself. Our role is to encourage and enable both child and family to meet the situation with which they are faced in their own unique way, stepping in with advice or practical help only with the greatest degree of sensitivity; any intervention should be invited, never imposed. Our overall objective should be to help the children and their families achieve the best quality of life, physical, emotional, and spiritual, throughout the lifetime of the child, and to help the survivors to live on after the death of their child.

It can be seen from this that the popular concept of 'hospice' as 'a place where you go to die' is severely misleading and limiting. The original meaning of the word was a place of rest and refreshment for those on a journey and, while this definition can be linked to modern hospice care, we must not equate hospice with a building. Hospice is a philosophy rather than a facility, a whole approach rather than an inpatient unit. This approach can be put into practice wherever there is a child who suffers from an illness which is a threat to survival, whether the child is at home, in hospital, in residential care, in school, or in a hospice building.

The family should have a choice of places and services during the different stages of the illness. If they choose to be somewhere other than home from time to time, they should also be allowed to decide whether they will accompany their child and, if they do so, how much of the caring for the child they themselves continue to do. Most parents, initially at least, wish to take 100 per cent of the responsibility of looking after their sick child themselves. We do not help by trying to force them to do otherwise, even if it appears to us that the parents need an opportunity to rest and build up their reserves. If we allow time for a trusting relationship to develop between the child and family on the one hand and the professional caregivers on the other, the family may eventually be ready to share the care of their child without feelings of guilt or loss of control.

It follows that it should be possible for the family to choose where their child will die. Occasionally it will be the child's own choice. What matters is that it is the place where both child and family feel at peace, in control, safe, and secure; and for some, when actually faced with the child's terminal symptoms, this may not always be in their own home.

If we are to ensure the greatest good for the family then we must do our utmost to keep lines of communication open with all the other health care agencies involved, co-operating with them in any way which will be of benefit to the family, remembering that this applies during the lifetime of the child and over the undetermined period of bereavement following the child's death.

Families will experience exhaustion affecting every level of their being; loneliness when friends and relatives are unwilling to be involved because of their own feelings of inadequacy in the face of such tragedy; and emptiness drained by grief in all its endless energy-sapping manifestations. Parents will often feel that they are going mad, so terrible is the pain of their experience. We cannot take away the cause of their pain, nor even answer many of their questions, but we can offer to be there alongside them. We should be there not just as professional caregivers but as fellow human beings and trusted friends with whom there need be no pretence.

The keynote to paediatric care must be flexibility. A 'take-it-or-leave-it package' to use the words of one father, is of little benefit to the families of children with life-threatening disease, for the needs of each family and each day are different. We can only begin to meet their needs by listening to and learning from the children and the families themselves.

Part of Me

I can't feel it,
I can't see it,
It's just a part of me,
They call it a tumour,
What a name,
No wonder it wants revenge.

It gives off no pain,
Or shows any discomfort,
Or even tells us where it is,
But slowly it's growing,
Slowly taking in my life.

It's just there,
To see on a screen,
It's overcome battles,
It's overcome pain,
But it's still there,
A part of me.

Why does it not show itself,
To prove to all that it is there,
But it just stays quiet,
A part of me.

Tracy Wollington

'I'm still running'

TRACY WOLLINGTON

Tracy Wollington was nine years old when she was found to have a lymphoma. She underwent two years of chemotherapy after which she remained well for another three. Then she developed another totally different form of cancer. It was too far advanced to be surgically removed and failed to respond to either radiotherapy or intensive chemotherapy. For the last three years of her life, Tracy lived with the knowledge that she had an incurable illness. She was helped to come to terms with this awful truth by writing.

I'm still running is a book of poems written by Tracy and published shortly after her death. Some of the poems reflect her anger at confronting serious illness, death, and grief; however, when she was terminally ill she had conquered her anger and was at peace. Proceeds from the sale of the book go to help CLIC Trust and St Luke's Hospice, Plymouth.

She bore her illness with the most remarkable courage and dignity. She died at the beginning of August 1988. She was 17 years old.

John Halliday

Wollington, T. (1988). *I'm still running*. Pub. J. J. Halliday, Winsford, Totnes Road, South Brent, South Devon.

Part II

Personal reflections and domiciliary care

2 The Cancer and Leukaemia in Childhood Trust (CLIC)

ROBERT WOODWARD

The Cancer and Leukaemia in Childhood Trust (CLIC) was born on the 16 January 1976. My son Robert was born in 1966. He was always a very fit, athletic boy but quite unexpectedly he became ill at the end of 1973. Following six to eight weeks of ups and downs he was admitted to the Children's Hospital, Bristol. Early in the New Year my wife and I were called in to be told the news that he had a ghastly malignant tumour—a neuroblastoma. The dreadful reality came home to me when I was told that without treatment Robert would probably die within six weeks. It seemed unbelievable that this promising young life could be under such a mortal threat.

We were told that there was a doctor at the Bristol Children's Hospital who was collaborating with American colleagues on innovative treatment of children's cancers and had begun treating children at the Children's Hospital in Bristol with some measure of success. He proved to be young, enthusiastic, frank, and yet a very sensitive doctor who explained the difficulties and the seriousness of the situation. He outlined the horrific programme of chemotherapy, radiotherapy, and surgery, explaining the side effects that would occur during the two years of treatment. My wife and I were physically, mentally, and spiritually drained. But there was yet another blow to come. We were told that the initiation of Robert's programme of treatment coincided with the doctor's departure for a two year Fellowship in the USA.

No doubt his colleagues, who would be holding the fort in Bristol, were very capable. However, I felt the need for the ongoing treatment to be supervised by this young doctor. I recall a couple of days later driving to my office in Gloucestershire and suddenly having an idea of how the problem might be overcome. I came off the motorway and turned straight back to

Bristol. I sought out the Professor of Child Health at the Bristol Children's Hospital and asked him if it would be permissible to set up a travelling fund to bring the young oncologist back to England on a regular basis over the two year period. The plan proved to be acceptable to all concerned as a way of maintaining continuity and expertise with Robert's care. The travelling fund was duly set up and a two-monthly specialist shuttle established.

This initiative led to discussions on what was required to set up a first-class service for children with cancer in the south west of England. I was very grateful that we lived so near to the Bristol Children's Hospital, but realized that there were many parents in the West Country who were less fortunately placed. Many families travelled long distances to and from the treatment centre, in some cases more than two hundred miles—each way— from their homes to the hospital. This often meant that both parents could not leave the home base; usually it was the mother who came from far away with the sick child and stayed in the hospital, maybe sleeping in the ward, in the corridors, or in bed and breakfast accommodation. Father would be left behind trying to keep the home together, looking after the rest of the family, and holding down his job. It was apparent just how much additional heartbreak such problems gave these families. I felt convinced that the provision of family accommodation would reduce some of these pressures.

In 1974, after Robert's treatment regimen was established and with the travelling fund under way, I suggested to the Professor of Child Health that there was something practical I could do immediately. There was a bungalow in our village just outside Bristol, which belonged to my brother and myself (through our building partnership). We had intended to demolish it; instead we decorated and furnished it as accommodation for parents travelling to Bristol for their children's cancer therapy. For two to three years my wife, Judy looked after the visiting parents. Some families—both parents, and at times other children too— came and stayed for five to six weeks at a time. We saw first hand the tremendous difference it made to families in those circumstances. Little did I realize that by creating this 'home-from-home' we were pioneering an aspect of care in the United Kingdom.

The next priority in building a first-class paediatric oncology service was somehow to create a new consultancy post to head a regional service. However, even in 1974 there were such severe financial constraints that such a development was near impossible. Nevertheless, we set about the task of establishing a trust to create a level of excellence for the care of children suffering from cancer in the south west of England. The needs fell into three categories: Treatment, Welfare, and Research.

Treatment

We felt that the top priority was to ensure that any child diagnosed with cancer, of any sort, in the south west of England received the best possible treatment. To achieve this we would need to set up a consultancy post together with a team of domiciliary care nurses to ease the terrible strain of the repeated trips that children had to make to the oncology centre at the Children's Hospital. We realized there would certainly be many other things to do in the area of treatment, but these new posts took precedence.

Welfare

It was obvious that accommodation was needed: as near to the hospital as possible and in a property that would 'lift' parents' morale when they came; a place providing a welcome in secure and comfortable surroundings; a clean and homely house where if a child was well enough, the family could be united, meet other families (if they chose), and still have the patient attending the oncology centre daily. I felt it was important to keep the families together. There must be room for Mum, Dad, patient when well enough, brothers and sisters, and other family and friends if necessary.

Research

Thirdly, we would need to look at research to ensure that our team was in touch with and contributing to the frontiers of science in children's cancer studies.

In those early days, all these things were tantamount to building skyscrapers while not owning a wheelbarrow or a pile of bricks! But with the Professor's infectious enthusiasm and vision, I began to see the whole thing developing and building in my mind. It was very similar, I thought, to when some years previously in 1962, I had looked at a twenty-acre site. Then, I did not see the fields that other people saw, I saw a development of beautiful houses laid out with landscaping and flowers, mellowed stone work and mature buildings; I saw a whole community of people. By now, in 1975, that project was almost fully completed; fields had become houses, houses had become homes, and homes a neighbourhood. I could see the things that the Professor and I had talked about; and I could see that one day they could be achieved.

But we were back to square one. The immediate item was the consultancy post. The travelling fund was working well. Bristol University agreed that we could have the paediatric oncology post set up at the Bristol Children's Hospital providing we could put some money 'up front'. The amount of money requested was £140 000, a huge sum of money (even by today's standards!). We had no idea how we were going to raise such a figure. We invited the Director of the national Leukaemia Research Fund (see Appendix) to come to Bristol to discuss the problem with us. The upshot of his visit was that the Leukaemia Research Fund offered to pay half of the cost necessary to establish this post. The door was unlocked! We then found a private donor to covenant the other half of the cost. So we had the consultancy post funded ready for the beginning of year, October 1976. This was duly advertised in August 1976 and won in open competition by our oncology specialist who had returned from California. Tragically, this coincided with Robert's death in July of that year.

Establishment of the Trust

We set about establishing a trust to build a first-class regional service for children with cancer and their families. We were virtually ready to launch but we could not agree on a name for the trust. I did not want to restrict it in any way or confine it to one family name. I wanted it to be a trust that would grow and one day be able to embrace all afflicted families; I wanted all such families to be able to identify with it: as a ray of hope to a family whose child was being treated; or as a memorial for a family who had lost their child; or as a thanksgiving for a family whose child had come through and was cured. Then one day, it clicked! Within 24 hours I had named the Trust the Cancer and Leukaemia in Childhood Trust—(CLIC); within another 24 hours I had the artwork for our logo completed: we were in business! In 1976 we raised £3000; in 1977 we raised £8000.

In 1979, we took the huge step of starting our branch organizations. Branches were set up in Gloucester and Bristol by parents who expressed their wish to 'give something back' to the society which had cared so much for their afflicted children. (CLIC now has twenty branches in the south west of England which I believe, provide a good, even coverage of the region).

Our income increased and in 1981 we closed the bungalow and purchased Pembroke House in Bristol for parent accommodation, near to the Bristol Children's Hospital; we renamed it CLIC House and it also became our administrative headquarters. The 1980s have seen an expansion and consolidation of our activities on all fronts, never losing sight of the three pillars of the Trust's endeavours: Treatment and Research as well as Welfare.

Treatment

The Oncology Unit set up by CLIC in 1977 has treated more than 1100 children. Any child diagnosed in the south west of England with cancer can confidently expect to receive the best possible treatment on his or her doorstep. CLIC sponsors sessions in children's cancer clinics in Gloucester, Bristol, Bath, Taunton, Yeovil, Barnstaple, Exeter, Plymouth, and Truro.

We now have a team of nine domiciliary care nurses, each of whom is linked to one of the cancer clinics. Since our first domiciliary care nurse was appointed four years ago, the benefits for individual families have been evident. Parents tell us how much of the strain is alleviated by having their own specialist care nurse always available on whom to call any time of the day or night; someone able to come into the home as a recognized and accepted member of the family; someone to liaise with the school, explaining what might or might not happen; and someone to maintain the relationship between the general practitioner and the patient receiving specialist treatment.

CLIC funds a play therapist at the Bristol Children's Hospital and the secretary to the Malcolm Sargent Social Worker (see Appendix). We make personal grants in cases of extreme financial hardship and difficulty to individual families, always on recommendation from the relevant social worker.

Research

CLIC proudly boasts its own research laboratories in the Medical School at Bristol University. CLIC has also funded other major research into the immunology of cancer therapy in children jointly with researchers at the Bristol Maternity Hospital, the Neurosurgical Unit at Frenchay Hospital, the Bristol Eye Hospital (on retinoblastoma), and over the past two years CLIC has funded the tissue typing for bone marrow transplants, building up a register of donors at the UK Transplant Centre at Southmead Hospital.

We also fund two of our own CLIC Clinical Research Fellows; one linked to the Oncology Unit at the Bristol Children's Hospital, the other concerned with ACTs national survey of families' and carers' support services which forms the basis of their National Resource and Information Centre (see Appendix). We fund the computer in the unit; we also fund the operator. Our data base contains information on all the children who have been diagnosed and treated over the last fifteen years in the south west of England. We work closely with the newly opened Bone-marrow Transplant Unit located at the Bristol Children's Hospital, funded by the Regional Health Authority but built on the foundations and head of steam generated in the region by CLIC. In Gloucestershire, we have paid for the extension to the paediatric ward to give a day bed area, a small eight-bedded ward with kitchen and interview facilities attached to the Gloucester Royal United Hospital.

And so the work goes on with additions month by month to the fabric of comprehensive clinical care in the south west.

Welfare

We now have eight properties and the original CLIC House bought in 1981 has been more than doubled in size. We had the opportunity to buy the adjoining house four years ago and, by interlinking it with the original property, extended the house to accommodate children who were not able to cope with the stairs. CLIC House now comprises an eleven-bedroomed property where we can cope, fairly comfortably, with up to forty people at a time. Yet even now we find that with the Bone-marrow Transplant Unit coming into full operation we are having to turn people away. We realize that we shall not satisfy the demand unless we have a further extension. So we have recently purchased the house immediately behind CLIC House which will provide additional accommodation for at least another three to four families coming into Bristol at any one time.

Spinal and brain tumours are dealt with at the Neurosurgery Unit at Frenchay Hospital on the outskirts of Bristol. We have purchased and renovated an attractive cottage from the Health Authority which is situated in the Hospital grounds. We have also established a rest room and overnight stay facilities within the neurosurgery ward for those parents who need to be very close to their child at the most crucial stages of treatment. Children attending for neurosurgery in Plymouth are catered for at CLIC Lodge, a three-bedroomed attractive house within three minutes walking distance from the paediatric ward at Freedom Fields Hospital. Some of the paediatric work is now moving to the new Derriford Hospital, some five or six miles away; CLIC has therefore acquired a three-bedroomed flat—CLIC Court—within the grounds of Derriford Hospital.

CLIC also provides crisis-break accommodation for families when their knees are about to buckle, whether it be at diagnosis stage, or relapse, or when a child becomes terminal or when they have sadly lost their child, or whenever ... The paediatric social workers throughout the country have been notified of this facility. It comprises two crisis-break flats, one on the sea front at Sidmouth in East Devon, and another on the sea front at Weston-super-Mare. (Incidentally, three separate families travelled from as far away as Sheffield to stay at Sidmouth during 1988.)

One of the more recent facilities funded by CLIC is an ambulance service. We have two specially adapted Volvo estate cars. The CLIC ambulance provides comfortable transportation for the child who would otherwise travel with a great deal of discomfort (due perhaps to loss of weight, exhaustion, or pain) in a small private car or by public transport. We have a team of voluntary ambulance drivers from the St John's Ambulance Service and from the Avon Ambulance Service and someone is always available to ferry children in any direction, at any time.

We have established a regional liaison group entitled VOTE—Voice of the Experts—bringing together representatives of all disciplines involved in the care of children with oncology problems in the south west of England, to allow them the opportunity to advise CLIC on service improvements, wrong decisions, overlooked new specific areas of need, and on priorities for future expansion and development. A similar group has been established among the researchers receiving CLIC grants to provide regular reports to the Trust on the progress of work and for the researchers themselves to discuss the various works they are undertaking. These liaison groups ensure good relationships between the Trustees, carers, clinicians, and researchers throughout the whole of the south west region.

Future outlook/Further afield

CLIC is an army of people dedicated to the original aims of the organization. Our income last year was almost £700 000. I hope that in the 1990s we will raise at least £1 000 000 in the south west of England, regularly, on an annual basis. 1990 also sees the launch of CLIC UK, and we may now look to the other areas of the United Kingdom and ask if there is any way in which we can help, whether it is by advice, sharing our knowledge or experience, or by the provision of pump-priming funds to establish CLIC style properties, domiciliary care nurses, or any of the other facets in which CLIC has been involved and which are considered worthy of emulation.

CLIC is committed to establishing, maintaining, and enhancing an excellence of care for children and their families whose lives have been blighted by cancer and leukaemia, wherever they may be.

3 *Malignant disease in children: a view of a general practitioner and parent*

JOHN HALLIDAY

Malignant disease in children is rare. The average general practitioner in the whole of his career can expect to see no more than one or two cases. It could therefore be argued that it is scarcely worth considering such a condition from the general practitioner's point of view. Yet when a case of malignant disease occurs in a practice it can have a profound effect on a great many members of the child's family, also on the community as a whole. This is especially true if the child dies. There can be few more harrowing experiences for a family than to watch a previously healthy child die from cancer. The family doctor can play a major role in the care and support of both the child and his family.

When I became a general practitioner on the southern edge of Dartmoor in 1974 in a practice of barely 4000, malignant disease was certainly not uppermost in my mind: yet in the fifteen years since then we have had no less than seven children with one form of cancer or another.

1. *Renal carcinoma—three cases.* Two children have died, the third is alive and well several years after completing treatment.

2. *Acute leukaemia* This girl died only ten days after first becoming ill from spontaneous rupture of her spleen.

3. *Haemangiopericytoma* This little boy was diagnosed as having a large tumour on the left side of his neck when his mother went for a routine antenatal scan at about 24 weeks. He was delivered by Caesarian section at 36 weeks when the tumour was removed. Now, some three years later, he is alive and well.

4. *Non Hodgkins' lymphoma* followed three years later by a *fibrosarcoma.* This girl was nine years old when she was found to have a lymphoma. She underwent two years of chemotherapy,

at the end of which time all sign of her tumour had disappeared. Then three years after finishing treatment, she developed a totally different cancer—a fibrosarcoma in her abdomen. It was far too advanced for surgical removal and failed to respond to either radiotherapy or intensive chemotherapy. After a long drawn out illness, borne with great courage and the full knowledge of what was wrong with her, she died in August 1988 aged seventeen.

5. *Adrenal carcinoma* This little boy, Jamie, was my older son. At the age of five he was found to have a large tumour of his right adrenal gland. This was removed in London and for just over two years he remained fit and well. Then it became apparent that his symptoms were returning. Over a period of nearly six months he returned repeatedly to London for tests, investigations, endless X-rays, and scans; finally he underwent a laparotomy but the surgeon was unable to find any evidence of a tumour. After a further six months, he began to develop pain and it was found that his tumour had recurred and was infiltrating between his abdominal aorta and inferior vena cava. After a very extensive operation in Bristol the tumour was removed. Sadly, it returned again within only a few months. He again developed pain which rapidly became very severe. He underwent an intensive course of radiotherapy, followed a few weeks later by a spinal block. This totally relieved his pain but tragically left him paralysed from the waist down. He died peacefully at home a few weeks later in January 1980. He was nine and a half years old.

As a result of this experience I suppose it was inevitable that I should become involved in the care and support of children with malignant disease. In particular, since 1980, I have been associated with CLIC, the Cancer and Leukaemia in Childhood Trust (see Appendix) which has played such a large part in improving the care of children in the south west of England who develop malignant disease.

The parents' view

There is no easy way of telling parents that their child has a malignant disease. One moment their child clearly has a problem, the next moment the whole situation has escalated into quite a different dimension. Reactions run the whole gamut of human emotions—bewilderment, panic, a deep feeling of failure and guilt, even anger. There is a total sense of shock at the knowledge that suddenly the life of your child is threatened. Just as you are trying to grapple with this new situation, you are often informed that you are being immediately transferred to the regional paediatric oncology unit—this may be many miles away.

So a child and his family embark upon a routine of regular trips to the hospital, often for treatment such as chemotherapy. Always there is the nagging worry at the back of your mind about what might be found at each visit. The whole family has to learn to live each day one at a time. It becomes very difficult to plan ahead or to book holidays a few months in advance. It is rare that everything goes totally smoothly—there are inevitable alarms and crises, as a result of which holidays and trips have to be cancelled. Consequently, one stops planning ahead for fear of disappointment. The whole family is involved. The other children must be considered as their lives are greatly affected. Young children especially find it very difficult to understand why their mother is away so often with another brother or sister.

The extra travelling and staying away from home can prove to be very expensive. Telephone bills get bigger and bigger as you try to keep other members of the family abreast of what is happening. All the time you try to keep normal family life ticking over. The father has to continue working and one tries to present as normal a face to the outside world as possible.

It has been said that the rate of marital breakdown in a family with a child with malignant disease is far above the normal. I firmly believe that one of the primary causes of this is the inevitable separation of family and the lack of proper family accommodation for them close to children's hospitals where they can stay whilst their child is receiving treatment. It is one thing to have a child in hospital with an acute illness—we can all cope with a week or two of turmoil. It is quite another when a child is under long-term treatment. The return to hospital for the next course of chemotherapy is relentless—it can sometimes go on for years.

When we travelled to London with Jamie we had to find

accommodation wherever we could. One became adept at sleeping on ward floors—usually this meant we could not settle down
until after midnight and often we had to be up again between
5 and 6 am. It was pretty demoralizing and exhausting. When
we became involved in CLIC and started a branch in Plymouth
all of us in the branch felt our first task was to raise the money
for a house like the one in Bristol, close to the children's wards
where families could stay free of charge for as long as their child
is in hospital. This house was opened over five years ago and
has been occupied by one or two families virtually ever since.

One also has to cope with, and adapt to, the slow insensitive
routine at the hospital. We really did have to wait for hours,
not once but many times, in out-patients, laboratories or X-ray
departments. Often we would sit uncomfortably in corridors
waiting for one test or another whilst the hospital staff scurried
by, avoiding our gaze. I would sit there wondering why this
had to happen and hoping that some day, someone would find
a better way of organizing things. We knew that the level of
medical expertise to which we had access was as high as could
be found anywhere in the world, but there was no doubt that
the endless waiting in no way helped to boost our flagging
morale.

I believe that one of the big problems is that often hospital
staff may not be aware of the distances patients have had to
travel. The hospital itself is such an enclosed world that some
of those who work in it tend to forget the real world outside.
This was brought home vividly to me one day when we returned
yet again to London with Jamie for another scan. We had been
told very firmly that we had to be on the ward by 8 am so
we made the two hundred mile journey the day before, stayed
with friends overnight and dashed across London in the early
morning, arriving on the ward just on time at five minutes to
8. We were met by the staff nurse who asked us why we were
there. We told her. 'Oh! Haven't they rung you?' she said, 'The
machine isn't working this week.'

The pastoral care of children with malignant disease, and
indeed all the life-threatening diseases, is improving. I am convinced that the only reason it seems to have been lacking at
times is because there simply has not been the personnel available
to cope with the work load. Medicine has changed dramatically

in recent years, especially in fields such as paediatric oncology. The treatment is so long and drawn out that the old, traditional forms of medical and nursing care must change in order to help provide support for those who have embarked upon treatment. Today, the majority of children with malignant disease are cured. However, even when treatment has finished families still require close support—minor infections and fluctuations in a child's health carry a much greater significance for a family whose child has been treated for cancer.

For some families the treatment is not successful. The child and his family have to be prepared for a terminal illness. Increasingly, today, the child will remain at home, close to the family. I believe that in most cases this is the appropriate place for the child to be. When my own son was dying we were deeply grateful for the support of people who had already become familiar friends both to Jamie and the rest of the family. Our paediatrician, health visitor, and district nurse all gave tremendous support and helped to guide us through these last weeks. Now, nearly ten years later, my own memories of that time are not without remembrance of laughter and considerable humour, as well as deep sadness.

A general practitioner's view

I believe that the general practitioner has an important part to play in the care of a child who has malignant disease. Unfortunately there are a number of factors which can arise and which sometimes confuse the situation. Since I became involved with the Cancer and Leukaemia in Childhood Trust, I have met many parents who have, or have had, children with one form of malignancy or another. In talking to them, it is quite apparent that the role of the general practitioner is somewhat ambiguous. Many families have been full of praise for their general practitioner, but others have been angry, believing that he was not interested in their child's illness and their own situation.

There are a number of problems which seem to arise with regularity and these need to be recognized so that they can be dealt with efficiently. In order to do this, I have looked at the role and involvement of the general practitioner at three stages:

(1) initial diagnosis;

(2) successful therapy;

(3) unsuccessful therapy and terminal care.

Initial diagnosis

Mercifully, malignant disease in children is not common. This does not help the general practitioner. It is most unusual for the general practitioner to make the diagnosis after seeing the child only once. Malignancy is not, nor should it be, the first thing that comes into a general practitioner's mind when he first sees a child with what might appear to be a minor complaint. Malignant disease can present in an enormous number of different ways—vague ill health, a niggling pain, a limp. I have heard many stories from parents who have taken their child to their general practitioner on several occasions when nothing seems to have been found, nor action taken. Sometimes several weeks, even months have passed before the diagnosis has been made. Sadly, when this happens the general practitioner is regarded as incompetent by the family. One hears the complaint, 'If only he had made the diagnosis sooner'. It is a difficult situation for the general practitioner and can put great strain on his relationship with the child and his family from the start. To prevent this happening there are, I believe, some fairly basic rules of practice.

1. Always be prepared to see a child.

2. If you are unable to find any abnormality after examination always tell the parents you cannot find anything, but are prepared to examine the child again if symptoms persist.

3. Always take seriously the mother who comes to the surgery with her child and tells you that while she does not know what is wrong, she knows her child is not right.

4. Beware of telling a family categorically that there is nothing wrong with their child.

5. Note how often the child is seen. If after a few visits you have found nothing, consider asking a general practitioner

colleague or a paediatrician to see the child. A new pair of eyes may spot something you have missed.

As doctors, we sometimes dismiss as ridiculous symptoms which patients describe to us. They may sound bizarre but in fact may be quite genuine and outside our experience. At one stage of his illness, my son Jamie took to watching television whilst almost literally standing on his head. He looked quite extraordinary and would push his feet as far up the wall or the back of the chair as he could and view in an upside down position. When a short time later we found his tumour had recurred and was causing obstruction of his inferior vena cava the explanation of his behaviour became apparent.

The diagnosis of malignancy is usually made in hospital. I believe that it is of great help to the relationship between the general practitioner and the family if he can take the trouble to actually visit the child in hospital. Patients tend to compartmentalize their doctors—it helps enormously for them to see their family doctor in the hospital context, actually communicating with the consultant and his team. At times, it may be inconvenient but it is well worth the effort.

In the last two years CLIC has set up a team of domiciliary care nurses in the south west. (Their role is described in Chapter 4.) One of the most important features of their task is to contact the primary care team as soon as the diagnosis has been made and to brief them on what, for the vast majority of primary health care workers, is a very unfamiliar situation. The domiciliary care nurse will have met the family and, hopefully, will be able to discuss whether there is an antagonism towards the general practitioner. Should this be the case, then once recognized, the rift can usually be repaired.

Successful treatment

A general practitioner's role and involvement can be very limited in the case of the child and his family for whom treatment is going well and is eventually successful. Some would question whether he should be involved at all. I believe he should. There is a natural tendency for the general practitioner to be somewhat diffident and feel out of his depth when he is confronted by the complexity and intricacies of paediatric oncology. This is inappropriate. There really is no need for the general practitioner to be fully clued up on the minutiae of chemotherapy. The role of the general practitioner is to be more of a sounding board and a support for the family and their child. Some oncology centres send out information packs to enable the general practitioner to obtain a broad view of a child's complaint. With this background the general practitioner can deal with the queries and doubts which do not need the specialist to answer them. Moreover, there is always a need to keep an eye on the rest of the family in case the other children are suffering, or there is marital stress.

Sometimes treatment is not successful and perhaps, over a period of years, the child becomes terminally ill. It does not help if the general practitioner has scarcely been seen all those years, only reappearing as the child returns home, terminally ill.

The terminally ill child

If treatment fails, it is increasingly common today for every effort to be made for the child to be cared for at home. I firmly believe that in most cases this is the appropriate place for him or her to be. A general practitioner can play a full and active part in the care and support of a terminally ill child. It is a situation that will not be without problems and consequently benefits from some forward planning.

The domiciliary care nurse can be of vital help in liaising between the hospital and the community when it becomes apparent that a child is not going to respond to treatment. When this happens it is important that the members of the primary health care team—the general practitioner, district nurse, and health visitor—meet with the domiciliary care nurse, the paediatrician, and the Malcolm Sargent Social Worker (see Appendix), to plan a scheme of care.

As far as the general practitioner is concerned one member of the practice should assume overall responsibility. But he must also involve one or two colleagues from the practice who can cover for him from time to time. It is not practical for one person to assume sole responsibility for a child, it must be shared—this applies as much to doctors as to nurses. There will be those who will assume more of the load than others but it is vital that *all* the carers meet regularly, share problems, and support each other through what can be a harrowing experience. It is important to recognize that in the community where fewer people overall will be involved than in hospital, the stresses and strains may, consequently, be greater.

The general practitioners involved with a terminally ill child in their practice should recognize that some of the arrangements for out of hours cover in many practices may not always be ideal. It is not unusual for general practitioners to participate in duty rotas which can involve quite large numbers of doctors. For a terminally ill child, the out of hours cover should not involve more than three or four doctors at most. The last thing a family with a dying child wants, having sent for a doctor in the small hours of the morning, is to be confronted by someone they have never met and who knows nothing about their child. Familiarity and trust between all members of the team is a vital requirement for the child and his family as he enters his last few weeks.

The domiciliary care nurse brings to the community a breadth of experience and expertise which is beyond that of the primary health care team. With her help and advice and the judicious use of all the modern aids that are now available, the care of a dying child at home should be a practical proposition for most general practitioners. Help can also be elicited from the local hospital but I would submit that this is more likely to happen in the large towns; in more remote rural areas it may be too far for the hospital-based paediatric team to offer practical help other than advice.

Children rarely die from cancer in this country today. I believe that, at times, modern medicine prevents us from recognizing that terminal care requires a special commitment from those of us involved with such children and their families. The general practitioner will not be confronted by such a situation very often. When it does arise, it is of value to all concerned for the general practitioner to be fully involved sharing with a family an experience which neither he nor they will ever forget.

4 Domiciliary nursing care
SALLY CURNICK

In the UK, 1 in 10 000 children aged from 0 to 14 develop cancer each year. Within the South Western Regional Health Authority up to 80 new patients and 25 to 30 deaths are expected each year. The Cancer and Leukaemia in Childhood Trust— CLIC (see Appendix) is supporting a team of nine domiciliary care nurses throughout the south west to work with families whose children have been afflicted by a malignancy. The regional Oncology Unit is in the Bristol Children's Hospital, with local treatment centres at Gloucester, Bath, Yeovil, Barnstaple, Taunton, Exeter, Plymouth, and Truro. The majority of children are managed in Bristol initially for diagnosis and induction treatment; the maintenance treatment is usually continued at their local centre.

The CLIC domiciliary care nurse is able to meet the children and their families whilst they are still in hospital and then visit them at home following discharge to provide ongoing support. Parents and children find their home surroundings far more conducive to relaxed discussion about their feelings, fears, and hopes, than the clinical atmosphere of the hospital. The value of keeping the child out of hospital as much as possible cannot be over-estimated; by supporting the patients and their families, both physically and emotionally, clinic time and hospital admissions can be reduced. Moreover, this allows siblings, grandparents, and other important members of the family to be included in discussions, thereby harnessing their input into the supportive unit of the family.

Because management of children with cancer is highly specialized and rare, general practitioners can have little expert knowledge of the subject. The initial symptoms are often vague, leading to a delay in diagnosis; this can cause a loss of confidence in the general practitioner. In addition, the intensive treatment at the hospital results in a close association being formed with the hospital staff during a traumatic time. It is especially important, therefore, that the family sees the hospital and the general practitioner working together, to restore their confidence in the primary health care team for the future. The domiciliary care nurse has an important role in liaising with the general practitioner and the primary health care and hospital teams. Her work encompasses both the educational and social needs of the individual child and family.

The physical support at home will include:

- routine blood counts;
- care of Hickman catheters;
- mouth care;
- eye care;
- changing dressings;
- removal of sutures.

The possibilities of giving some intravenous chemotherapy and blood products are being explored.

Schoolchildren are encouraged to return to school as soon as possible and the nurse is able, together with the social worker to liaise with the teachers, with parents' permission, to ease the transition back into school. The physical changes of the illness —alopecia, weight loss, insertion of a Hickman catheter, or a scar, can make it very difficult for the child to face school. The nurse can maintain contact with teachers throughout treatment, and when necessary, in terminal care and bereavement follow-up.

For terminal care, the objective of my role, and that of the few others like it around the country, is to allow parents faced with caring for a dying child to make a real choice between hospital and home care, and to have the necessary professional support to permit them to carry out that decision.

In the event of treatment failure, a decision is made with the family that active anti-cancer treatment is no longer appropriate and the emphasis of care changes to relief of physical and mental distress. The aim is to give the child as good a quality of life as is possible in the time left to him.

When families opt for home terminal care, they still retain their links with the hospital. The consultant will nominate a

clinical assistant, myself and the social worker already known
to the family, to work with the general practitioner and his team.
Together we can offer twenty-four hour support, pain and symptom control, guidance on how to communicate with the dying
child and/or siblings, emphasis on living what is left of that life
to the full, and eventually a 'good' death. The 'out-of-hour'
calls do not fall exclusively on the general practitioner; hospital
staff often give families their own home telephone numbers for
continued contact. Every effort is made to cater for the wishes
and comfort of the child and his family within their environment.
The general practitioner needs the additional support of the
hospital team, especially if he is single handedly acting as twenty-
four hour carer, in addition to his normal duties.

All possible distressing symptoms are considered in the light
of the knowledge of the natural history of the disease, and ways
of preventing or treating them considered. A plan of joint care
is arranged, tailored to the desired input of the primary health
care team. One or two members undertake to maintain regular
contact with the family, ensuring the continuity necessary to
enable support and counselling to develop over the terminal
phase. The relationship developed in this way enables contact
and counselling to be continued after the death of the child
for a period to suit individual needs.

Honesty is regarded as the appropriate policy from the time
of diagnosis, so if the child is old enough, and with the parents'
agreement, the subject of death and dying might well be discussed. If parents need help to explain death to their children,
I suggest two books, *Waterbugs and dragonflies—explaining death
to children* (Stickney 1984) and *The Dougy letter—(Letter to a
child with cancer)* (Kubler-Ross 1979) both write analogously
of life and death, using different forms of imagery. Most children
develop their own ideas on death from quite an early age, from
experiences such as losing an aged relative or pet, or from school
friends, or television. They do not seem to fear death in the
same way as many adults.

One teenager was greatly relieved when he remembered he knew
someone who had already died. This 'auntie' knew just how he liked
his eggs because she had looked after him and his sister whilst their
mother was in hospital for an operation. Another 6-year-old recently
asked if there was anything to do up in heaven, 'You know, games

and things?' He seemed quite satisfied after discussing what possible games there might be there.

Many parents will never have experienced death in a close relation before and need to be encouraged to talk about their fears and fantasies of seeing their child dying, and the probable mode of death needs to be explained and discussed at length. A contact number should always be available and an open invitation to return to the hospital at any time should they so wish.

Life needs to be as normal as possible, which might include attending school or nursery part-time, the visit of a domiciliary school teacher, and having friends in to play, otherwise the child may feel isolated and become frustrated or demanding and aggressive. Stories and videos often become a normal pastime whilst undergoing treatment for cancer, and these are very useful for occupying some of the waking hours when concentration is at a minimum. Siblings feel more secure if their life is as routine as possible, it gives schoolchildren their peer support, and an escape route from their fears, anger, and jealousy over the sick child, and the attention he attracts. Siblings should be encouraged to 'help' if they want to and should not be excluded from the focus of attention. If they are old enough, it is advisable to give them a choice if they want to be present when the child dies, if they want to view the body afterwards, and/or attend the funeral.

A wide variety of home aids are available to promote the child's comfort and facilitate the provision of care by the parents. These include:

- mouthcare packs;
- incontinence aids;
- sheepskins;
- special mattresses;
- syringe drivers;
- suction apparatus;
- an intercom;
- wheelchairs;
- major buggies.

Night-time can seem very long and frightening, and to have a Marie Curie night nurse (see Appendix) to stay with the family can be very reassuring. This allows the family to sleep, knowing that someone is awake and watching over their child, ready to wake them if they are needed.

After death, time needs to be set aside to spend with the child's body, perhaps putting chosen clothes on, or just holding and caressing until the family, parents, and siblings are ready to call the undertaker. Funeral arrangements might have already been thought about, or not, depending on the family. Some teenagers think very carefully about their death, and like to make a will and divide some of their personal things among their family and friends. They may also express a preference for either burial or cremation, and sometimes choose favourite hymns for the funeral service.

Regular contact is most important immediately following a death, or else the family will feel very isolated from all the support they have had during the terminal illness. As with all previous care, bereavement care needs to be individually tailored to suit each family. An added pressure for parents might be that the siblings' grief may differ in comparison to their own, surviving children may resent prolonged grief of their parents, making them, the surviving children, feel insignificant. The first anniversary of a child's death is an important milestone, and most parents appreciate an acknowledgement that their child is not forgotten. Likewise, during the first few years there are other important dates such as the date of diagnosis, and of course, birthdays and Christmas.

The consultants offer each bereaved family an appointment, to be taken up when the family feels they would like it, to deal with unanswered questions on any aspect of medical care. The period of bereavement goes on long after this first year.

Since January 1988, the Malcolm Sargent Social Worker (see Appendix) and I have been holding monthly informal meetings for bereaved parents. Some parents have felt that after a time, when the world goes about its business, and others' lives have gone back to 'normal', it becomes no longer acceptable to talk about their dead child, and how they feel. The group offers them a safe environment to express their emotions amongst people most likely to understand other bereaved parents.

This year we have started some groups for siblings. One is aimed at siblings who have a brother or sister on treatment, and another aimed at siblings who have had a brother or sister for whom treatment was not successful, and who died as a result of malignant disease.

Families may need ongoing support for several years; many parents have said, 'You never get over the death of a child, but in time the pain of that loss becomes more bearable'.

References

Stickney, D. (1984). *Waterbugs and dragonflies—explaining death to children*. Mowbray, Oxford.
Kubler-Ross, E. (1979). *The Dougy letter—(Letter to a child with cancer)*. Friends of Shanti Nilaya UK, London (see Appendix).

5 The Rainbow Trust: a domiciliary crisis service
BERNADETTE CLEARY

Rainbow Trust is a charity which specializes in offering help at home during a time of crisis to families who have a child with a life-threatening illness.

The start of Rainbow Trust

The Trust grew out of my personal contact in November 1981 with a mother whose 12-year-old daughter, Rachel, was terminally ill and who wanted to come home from hospital for the last time—to die in her own bed with her own familiar objects around her. Her mother, Maureen, had no relatives to support her and could find no agency to help fulfil Rachel's wish. I was asked by an acquaintance of Maureen's to call in and see if I could help. I was able to give both practical and emotional support, making it possible for Rachel to stay at home. The quality of Rachel's remaining life became such that a very special relationship formed between her mother and myself. After her daughter's death in May 1982, Maureen told other families with terminally ill children about the support I had given them.

Inescapably, I was asked to help families cope when pressures grew, and I continued in this way until it became evident that there were far more families seeking this kind of flexible support than one person was able to give. In order to respond to all these requests, to join these families at home in times of crisis, to become an 'extra parent' to the children, and a practical 'friend' to the parent(s), a team of domiciliary workers was needed, each member of which could become this extra caring 'companion'. To this end, in September 1986, Rainbow Trust was founded.

The Rainbow Trust domiciliary team

Our domiciliary team members give round-the-clock support to parents, sharing and supplementing the expertise and skills parents need to sustain their families and themselves, practically and emotionally, organizing the household, caring for the sick child and siblings, and supporting the adults. The principle behind our commitment is 'flexibility'. Consequently, our workers, chosen with this as their prime attribute, have come from varied backgrounds. The current team comprises a sick children's nurse, a psychiatric nurse, and two nuns, one of whom has worked in the hospice for the elderly members of her Order, the other who has served her Order with all the 'housewifely' skills. We are soon to be joined by a paediatric nurse and a former Social Services family aide worker. It is important that the support given is tailored to suit each family, since each is unique in its way of coping with the traumas surrounding serious illness.

Despite some of the team having nursing qualifications, they do not offer nursing care as such. Their brief is to provide the basic caring skills which in normal circumstances, parents would themselves provide. They are, however, particularly well placed to liaise with a family's medical and nursing team, when, for example, a catheter needs unblocking or a child is in pain.

Our domiciliary workers help the family with domestic work, washing, ironing, cooking, shopping, babysitting—sometimes with the sick child, other times with siblings, often arriving to care for one, to be asked instead to take charge of the other. Involvement with all the family may well include taking children on outings or picnics, swimming, rota duties, shuttles to and from schools and clubs, having friends in for tea, and indeed, helping in any way to maintain an ordinary family life. If the sick child is re-admitted to hospital, along with Dad (or Mum), we keep the home running as smoothly as possible. This enables Mum (or Dad) to remain in hospital with their sick child.

Listening is part of our befriending. Each Rainbow Trust worker is given counselling training, either by an accredited British Association for Counselling teacher (see Appendix) or a former training counsellor from Relate (previously Marriage Guidance). They are subsequently supervised by their tutors when they begin to work with families.

Funding and referrals

Rainbow Trust is a registered charity and offers its services free to the families it supports. Whilst some of our funding comes from businesses and grant-making trusts, with one domiciliary

Table 5.1 Case information for twelve months

Case commenced	How referred	Nature of illness	Sex	Age	Geographical area	Case ended
4.7.88	Family	Neuroblastoma	Male	1½	Surrey	
24.8.88	Social worker	Metachromatic leucodystrophy	Male	6	Middlesex	April 1989
27.9.88	Social worker	Neuroblastoma	Male	7½	London SE19	April 1989
18.10.88	Social worker	Unidentified neurodegenerative disease	Female	9	London SW13	Jan. 1989
28.10.88	Head of school	Neuromuscular degenerative disease	Male & Female	6 5	Surrey	
24.11.88	Social worker	Acute myeloid leukaemia	Female	4	Kent	April 1989
18.1.89	Family	Ewing's sarcoma	Male	6	Herts	August 1989
27.1.89	Social worker	Neuroblastoma	Female	3	Suffolk	
14.2.89	Social worker	Leukaemia	Male	3	London W13	July 1989
24.2.89	Hospital symptom control team	Neuroblastoma	Male	5½	London W12	
18.6.89	Social worker	Acute lymphoblastic leukaemia	Male	12	Suffolk	
29.6.89	Social worker	Metachromatic leucodystrophy	Male & Female	5 2½	London W2	
12.7.89	Social worker	Leukaemia	Female	7	Surrey	

worker's salary financed by King Edward's Hospital Fund, the bulk of our money comes from the public.

Referrals come from a number of sources including doctors, hospital social workers, head teachers, clergy, and the families themselves. Based in the south, we work primarily within the southern Home Counties; sometimes called on to travel further afield, we will be able to do this more frequently as team numbers increase. Table 5.1 depicts the case load of one domiciliary worker covering a period of twelve months.

As can be seen from the table, we still have cases which do not yet have a 'case ended' date. This is because we are very willing to offer bereavement counselling and follow up, if this is what the family wants. Some families may wish to use another agency for such counselling, therefore part of our job is to find out what is available in their location.

Rainbow House

In 1990 we opened a holiday haven to which families may come for a short break. Set in the centre of a residential community on the fringe of the village of Great Bookham in Surrey, Rainbow House will give families the opportunity to meet with others in similar circumstances which, hopefully, will relieve the sense of isolation that some of them experience. It will be able to cater for three sick children and their families at a time, with a family suite on the ground floor and two parent-and-child rooms upstairs. The atmosphere will be very much one of a home-from-home, staffed to provide physical and emotional support 24 hours a day.

Summary

Rainbow Trust is based on my experience, since 1981, working with families where there is a terminally ill child. Such families need both holiday havens and extra pairs of hands in their homes. Our present domiciliary team works on the basis of those years of experience and the information amassed from the parents of those families helped so far; guided by the parents, we aim to give maximum support in the most effective way. Each family is unique and their needs vary; flexibility is the essential hallmark of our approach.

6 *Domiciliary symptom control*
ANN GOLDMAN

Background

In this county, there are about 1200 new cases of children with leukaemia and solid tumours presenting each year, and over 10 per cent of these (around 140) are treated at the Hospital for Sick Children, Great Ormond Street, London (GOS). The Haematology and Oncology department has for many years been concerned about both the physical and psychological care of these children and their families, and works as a team with social workers, play specialists, psychologists, and teachers, as well as medical and nursing staff. However, working as a paediatric oncologist in the department, it became clear to me that there were still a number of areas in which care could be improved and the Symptom Care Team was designed and developed to fulfil these needs.

Symptom care during treatment

For all children having treatment for cancer, there are side effects: nausea, vomiting, constipation, needle phobia, anxiety, and pain from procedures. Some families cope incredibly well, for others these problems are so intense as to make them consider refusing therapy in spite of being aware of the life-threatening nature of their disease. The nursing and more particularly medical staff are focused on treating the cancer itself and only see the family in the hospital environment; attention to these sorts of problem tends to be underestimated and takes second place. There was a need for someone to focus on and assess these aspects of care.

Terminal care

It has always been the policy and practice at GOS to encourage families to have as much choice and control as they want if their child is terminally ill. In the event, this has usually meant that children went home and local hospitals or primary health care teams took over the majority of care, with GOS maintaining contact by telephone. A number of children were readmitted terminally to GOS, some went to their local hospital and some died at home, receiving variable support.

The situation of having a terminally ill child with cancer at home is very uncommon and probably only occurs once or twice in the lifetime of most family doctors and infrequently even for general paediatricians. It is not surprising that it provokes considerable distress and feelings of inadequacy for everyone concerned. For the staff in GOS whose aim is to cure patients and where much of the care on the ward is 'high tech', it is emotionally difficult to make the change in gear to palliative care. It is also practically difficult to find the time and continuity of staff needed to care for a dying child. There seemed to be a need to provide a source of experience in dealing with both the psychosocial support and the symptoms of dying children and support for their families and the staff involved.

Liaison

GOS patients come from a very wide geographical area and there is severe pressure on beds and staff. One way this has been dealt with is to develop a system of shared care. This offers the family local-based care; but families often perceive, and are worried by, small differences in approach between the two centres. Communication and liaison could be improved by having a link person between GOS and the local hospitals.

Education and research

The study of symptom control and palliative care among adults has not been paralleled for children. Advice in either the hospice or paediatric oncology literature has been scanty. Although the principles of palliative care are similar for children, it is not enough merely to extrapolate from experience with adults. Research, study, and training in paediatric palliative care is much needed.

The Symptom Care Team

Having perceived the needs of the department and designed
a team which could fulfil them—and which the consultants in
Haematology and Oncology met with enthusiasm, the major
hurdle was to find funding. Eventually a private charity, The
Rupert Foundation, (see Appendix) dedicated specifically to the
care of children with life-threatening diseases undertook to fund
the work and the Team began to function in 1986.

The Team consists of myself, three nursing sisters all paedi-
atrically qualified and with backgrounds in oncology and community nursing, and a part-time secretary. It is hospital-based,
integrated with the existing members of the department and
extends care into the community. Each member of the Team
has a cumulative case load based on a geographical area. The
area we cover is essentially the North East, North West, and
South East Thames District Health Authorities, and we travel
to the patients. A 24 hour on-call system, both for telephone
advice and urgent home visits, is available for the patients and
their carers within the community.

Patients

The Team is involved with patients both during treatment and
terminally. There are about three newly diagnosed patients each
week and we plan to meet them towards the end of their first
admission. We arrange a visit shortly after discharge, which is
always an anxious time, and try to meet their community team.
We maintain contact with the patients on treatment, although
this tends to be reduced gradually unless a family has particular
problems. In addition, most weeks one or two children relapse,
then either we make or re-establish contact with these families.
Some children will go back on treatment again, and if so, we
remain in close contact with them. For others, the decision may
be that no further treatment is appropriate. This group, whom
we consider terminally ill, usually number between seven and
twelve at any one time. They tend to be the ones who take
up the majority of the Team's time. We also maintain contact
with the families of children who have died.

To illustrate the nature and scope of our work, I will refer to the group of children who died in 1987.

52 children from our department died in that year. This included 10 children on active treatment with the prospect of being cured, who died mainly in GOS from problems related to treatment. 42 children with progressive disease died mainly at home with some in the local hospital (Table 6.1). Of the 5 who died in GOS, 4 were still having some chemotherapy at the request of the family, and one died within 48 hours of relapse. Of the children in the local hospitals, 3 were still having some chemotherapy. The children at home were having palliative care only.

Table 6.1 Summary of deaths in 1987

	Total	GOS	Local hospital	Home
On treatment	10	7	2	1
Progressive disease	42	5	11	26

There were 16 boys and 26 girls in the group dying of progressive disease. They ranged in age from 4 months to 16 years (median 4 years). They had a range of diagnoses (Table 6.2). The time during which they were terminally ill ranged from 2 days to 13 months, with a median of 2 months.

Table 6.2 Range of diagnoses, 1987

Acute lymphoblastic leukaemia	14
Neuroblastoma	11
Acute myeloid leukaemia	5
Rhabdomyosarcoma	4
B cell lymphoma	2
Ewing's sarcoma	2
Histiocytosis	2
Hepatoblastoma	1
Juvenile chronic myeloid leukaemia	1

Management

The type of care the Team gives to the progressively ill children
includes: symptom management, liaison, and psychosocial support.

Symptom management:
This varies according to the needs of the child and the pattern
of spread of his particular tumour. We are involved with assessment of symptoms, planning, and then establishing treatment.
This is carried out in co-ordination with the parents and the
local primary health care teams. Pain is a great fear for parents
and does afflict most children at some time, especially those
with solid tumours. It is also an area of concern for community
teams who are often afraid of using the necessary strong analgesia
for children. A whole range of other symptoms may occur. Some
of the more common are gastro-intestinal problems (nausea,
vomiting, constipation, and sore mouth), bleeding in the leukae-
mic children, and convulsions.

We have a range of equipment available which can be loaned
out immediately such as syringe pumps, extra large pushchair
buggies and special mattresses and sheepskins to avoid pressure
sores.

Liaison:
Liaison between the various agencies involved in caring for the
terminally ill child is an important aspect. In this group of
patients in 1987, a wide range of different agencies were
involved:

- Hospital for Sick Children, Great Ormond Street
- local hospital
- radiotherapist
- general practitioner
- district nurse
- health visitor
- community paediatric nurses
- Cancer Relief Macmillan Nurses
- pharmacists
- schools, clergy, neighbourhood support, and self-help groups

Families vary in the extent to which they wish to handle the child's death themselves or do so with the help of others. Some families have had only one agency involved; at the other extreme, several families have had up to eight different ones. For most, three or four groups of care givers were involved. Sometimes we initiated their involvement and sometimes the family arranged it themselves. No standard pattern exists and for each family it depends on their own desires, the community help available in their area, and the personalities and past experience of the staff.

We spend a lot of time talking with, and listening to, families and the children; helping them face the emotional burden and also, anticipate the practical problems. The sort of questions which recur include, 'How long will my child live?' 'What shall we say to the child, to the brothers and sisters, to the rest of the family?' 'Will there be pain?' 'What can we do for it?' 'Who will do it?' 'Will we be able to manage at home?' 'What will the death be like?' 'Will you be with us at the death?' 'Is it terrible to plan the funeral before he dies?' 'What do you do after someone has died?'

Time and willingness to listen, honesty and flexibility to help each family handle their child's death in their own way is what we hope to be able to give.

In the first six months of 1987 we analysed the number of phone calls and visits we actually carried out. (Tables 6.3 and 6.4).

Table 6.3 Symptom Care Team: January–July 1987

198 visits	Patients	103
	General practitioners	32
	Local hospital	27
	Health visitors/district nurses	18
	Pharmacists	3
	Funerals	15

193 ward interviews
79 clinic interviews

Table 6.4   Symptom Care Team: January–July 1987

	Calls	
	Made by team	Received by team
Patients	320	114
Local hospitals	78	25
GP	67	15
Health visitors	64	29
District nurses	15	19
School	6	1
Total	550	203

Psychosocial support: When a child has died, quite often members of the team and GOS ward staff will attend the funeral and the Team maintains contact with the family. There is no formal 'aftercare' service but phone calls and visits continue to be made after bereavement for an indefinite period; again we have a flexible approach, attuned to a family's needs. Flowers are sent to the family about two months after a child has died from the whole Team at GOS and around this time an invitation is made for the family to attend a group meeting with other parents whose children have died. We hold this in conjunction with the social workers every three months. Families also regularly receive some form of communication on the anniversary of their child's death, on his/her birthday, and at Christmas.

Summary

The Symptom Care Team is part of the department of Haematology and Oncology at Great Ormond Street. We are involved with all the patients, from diagnosis, during treatment, and through terminal care and bereavement. The focus is on management of symptoms, liaison, and support at these times. For families with terminally ill children, we aim to help them choose where and how they would like to manage the death, and give them the help they need to achieve it, whether it is in hospital or, as most prefer, at home.

Dedication

These words were created
For special people.
These people are so special
To me.

Why does life
Make it so difficult
For me to say
'I love you'.

My mouth won't move
But my heart screams
As anger gets it's grip
I make you cry
Do you understand
How confused I am?

Perhaps one day it
Will be easy to say
But you are special people
I hope you realise
'I love you'.

I may not be an angel
Shout at you
Abuse your trust
Take your love
But who else has
Helped me so much.

You took my hand
While I slept
Held my head
So the pain is eased

Dedication

I am sorry to have caused you
So much trouble.

But always remember
No matter what anger states
My heart says
'I love you'.

Tracy Wollington

Part III

Children's respite hospice care

7 *Helen House*
EVE HERD

Sister Frances Dominica founded Helen House which has now been open for almost eight years. Helen is the name of a little girl, born in 1975. She was, in Sister Frances' words,

'... a lively, loving, healthy child of two and a half when suddenly she became ill and was found to have a cerebral tumour. Despite skilled surgery she is now in the same condition as she was soon after her operation—totally helpless, unable even to turn herself over in bed, unable to communicate with us in any way and scarcely aware of her surroundings.' (Dominica 1982)

Having got to know Helen and her parents well during her six months in hospital and realizing something of the strain they were under once Helen was at home, Sister Frances asked them if they would trust her enough to 'lend' her at times. This happened and she came to stay in her room at the Convent on many occasions, for short visits, so enabling her parents to have some unbroken night's sleep, the opportunity to take her little sisters to visit friends and relatives, and the chance to have a holiday.

Helen House is quite simply a special friendship with Helen and her family extended to other gravely ill children and their families.

Helen House was built in the grounds of All Saints Convent, Oxford, and is designed to be as much like home as possible. We want the children and their families to feel that they are coming to friends for a holiday. We do not wear uniforms but work together as a team trying all the time to create a family atmosphere. Our staff consists of nurses, nursery nurses, a teacher, a social worker, two part-time doctors, a physiotherapist, and other helpers. We have room for eight children at any one time; their parents and brothers and sisters are welcome to stay here too if they wish and many of them take advantage of this opportunity. Though there is an underlying framework to each day hinged on the washing, dressing, and feeding of the children, there are no fixed times which have to be adhered to. Rather, we try to learn from each family the pattern of care they have for their child at home and to follow their routine as closely as possible. This implies a need for great flexibility which may lead to chaos at times as healthy siblings play noisily among feeding trays or jump over their inactive brothers and sisters! It makes considerable demands on our staff and, to work well, needs a firm underlying structure of good communication, both verbal and written, between team members, and trust. So notes, files, booking sheets, and care sheets play an essential but largely hidden part in the smooth running of every day.

Looking back to the first year, 1982–83, Helen House opened quietly. 52 children came to us in that year, sometimes only two or three children staying at a time. They came from far and wide and varied in age from less than a year to 19 years. 25 of these had progressive, degenerative, life-threatening disorders such as Battens disease, Sanfilippo syndrome, and Duchenne muscular dystrophy. These diseases involve slow progressive deterioration over periods of years, leading to total dependency and eventual death. Of the others, 8 children had neoplastic conditions, and 19 had various congenital, metabolic, and non-progressive CNS disorders.

Numbers of children coming to Helen House have increased steadily each year. 75 children used us in 1985, 80 in 1986, 96 in 1987, 110 in 1988, 104 in 1989. Bed occupancy has risen from 43.8 per cent in 1983 to 60 per cent in 1989.

In 1989 we are currently caring for twice as many children as in the first year; approximately 60 per cent have progressive, degenerative diseases of genetic origin. Each case is a nightmare for the parents who, after the initial shock of the diagnosis, try to cope with the endless physical demands and come to terms with the heart-rending emotional implications of caring for a chronically ill child with a progressively life-limiting disease. Because of the relatively late appearance of symptoms of some progressive disorders (for example, Sanfilippo syndrome does not normally manifest itself until children are about 3 years of age) and because of their rarity, there may be delay in diagnosis; in addition quite often more than one child in the family is affected. We currently have thirteen families who have, or have had, two sick children and one with three.

So, many families who started coming to Helen House in the first year established a pattern of regular visits for planned respite; there were 220 admissions in 1982–83, the great majority of these for respite.

Sometimes there were emergencies at home and a family came when the crisis occurred and stayed until it was passed. It was decided that two beds should always be available for emergency admissions and though the majority of visits are for respite, those two beds are still kept for those in crisis. So Helen House with its colourful decoration and beautiful garden has come to be like a hotel—a good one we hope—offering care and entertainment appropriate to sick children, and rest and freedom to the other members of that child's family: freedom that is, to go shopping, or for a picnic, or to visit friends, outings rarely possible for many of our families to do together. It is very important that this includes healthy brothers and sisters who have so often missed out when their parents' attention is focused almost exclusively and quite understandably on their sick child. Not only can team members try to find time to listen to and play with them, but can also relieve their parents of the burden of care and of domestic chores, giving them the opportunity to pay sometimes overdue attention to their well children and, incidentally, to each other. So all members of a family may enjoy our jacuzzi, together, or with new found friends at Helen House.

We have room for two families in our two-bedroomed flat upstairs and parents staying with us are welcome to care for their sick child themselves, to share the care with us, or to leave it entirely to us once they have shown us how. For they are the experts in caring for their child and it is from them that we learn and build up our experience. We believe too that the care shared between ourselves and the families should be reflected in open communication with all the other services helping the family. Helen House is usually just one of a range of services the family may choose to use.

I would like to illustrate the role of Helen House in respite care by telling you about a family who have two children aged 8 and 5 suffering from Sanfilippo syndrome. They also have a healthy child aged 2. They have visited us since 1985 and usually stay for one week twice a year. This year, for the first time, the parents felt able to leave the two affected children on their own with us for two or three days of each visit. The father says this of their need for hospice support (Dominica, 1987*a*).

'We realise we can't grow away from our children or they from us. In fact, because of their increasing dependency the bond between us grows stronger and the responsibility towards Emma and William increases. This leads to phenomenal degrees of tiredness; no time for one another; little time to relax; social isolation and loneliness; feelings that you are on your own with your children; a knowledge that your friends have their life to lead and cannot help you very much; almost total self denial; feelings of guilt that you always fail in your attempts to do your best by all family members.

Traditional forms of help offered tend to emphasize the need to separate your child from you. Emma and William are a part of us—separation and institutionalized care are out of the question. The ethos of hospice acknowledges our family dynamics and deals with them where we are. We are not offered a take-it-or-leave-it care package, but one which understands our closeness to Emma and William and which meets our ability to 'let go'. The important thing for us is that the staff realise that they cannot offer us, the parents, any respite unless they can demonstrate that Emma and William are special people for them too. Helen House has become very much a second home. There, because of the love, we glimpse normality—the normality of sleep, physical and mental relaxation and being able to spend time with each other, recharging our marriage. The hospice is a place where people understand what it is like to care for a dying child. We don't get asked foolish questions about the illness; rather we are given information and support by people who are experts. We know that despite living 180 miles from Oxford the friends we have made there will help sustain us through the dark days which lie ahead. Emma and William's deaths will be a traumatic separation for us but we know that these friends will be grieving with us for they also love Emma and William. The hospice provides us with the best of all worlds, enabling us to play a part in our children's care while at the same time enabling us for a short while to put aside the worst excesses of caring for two very dependent children and engage in a little self-indulgence rather than self-denial.'

Six children died at Helen House during the first year. Since then 25 more have died here, 36 have died at home and 20 in hospital. We certainly feel that for most children home is the right place to die, surrounded by all their family and close friends and with all their familiar pets and belongings at hand, but for a variety of reasons it may not always be the choice of the family, and we respect their right to choose.

In her paper 'Reflections on death in childhood' (Dominica 1987*b*), Sister Frances tells of the support given to the mother of a baby boy who came to us aged 5 months with an inoperable congenital heart defect.

He was first admitted in August 1985 and died a month later, on a beautiful September day. His mother undertook most of his care for she knew she would not have him much longer. After his death his mother carried him into the garden, walked and sat with him in her arms for a couple of hours, often with one of us beside her, occasionally alone with him. She cried gently and talked to him. She remembered the day that he was born, also a beautiful sunny day: she talked of all the joy that he had brought into her life, and the pain. Then, in her own time, she carried him back into their room and slept for an hour. Then, and only then, was she ready to wash and dress him with very great love and care and without any sense of hurry. She chose the clothes he was to wear and the toys he was to have with him. She had already seen the room, furnished like a bedroom but able to be kept very cold, where he was to lie for the next few days. She carried him there. When his father arrived they visited him often in that small room, sometimes brushing his hair or rearranging his toys, sometimes lifting him out of his cot and sitting with him, uncurling his fingers and looking again at his hands, kissing him in the nape of his neck, lost in grief and wonder at the miracle that was their son. All the parents needed was our permission, spoken or unspoken, to do it their way, the way they knew instinctively, and our presence in the background and the knowledge that we felt pain and wonder too and were not afraid to show it.

We have our 'little room' as we call our mortuary, where parents, their relatives, and friends can spend as much time as they like with their beloved child. A member of the team is available to listen to the parents' ideas and plans for the time, place, and content of their child's funeral service if they so wish, and one or two of us who are particularly close to the family attend the funeral. Then life goes on with the usual bustle at Helen House.

Back home there are long painful days for the grieving parents where there is a yawning vacuum after all the comings and goings of doctors, social workers, and others who supported them before the child's death. They need the opportunity to express their grief; to cry and to talk of their loss and their pain—as they must if the grieving process is to happen naturally and to bring ultimate peace and wholeness to the sufferers. Their need for a sympathetic listener is as great after the death as before; indeed their pain may well intensify in the second year after the death, and loneliness is often a great problem. We keep in touch by home visits and by telephone. Three years ago, as life at Helen House became busier, Tessa, one of our team, started working as bereavement visitor and carries on the relationship with the families on our behalf. She travels great distances, in all directions, from her home in Coventry and devotes the time so desperately needed to listen to those who grieve. We welcome bereaved families back to Helen House when they feel ready to come. Each year we close for a short period (except for emergencies) and have two days—one used to be enough—when they are invited back for a few hours to see us again and, just as important, to meet each other. Many deep and lasting friendships are made between parents drawn together by sharing similar experiences of caring for their sick children—they understand the unique wavelength of each other's anxieties and problems. Sometimes Tessa tells us, usually after visits have continued for several years, that a family no longer needs support from us, that their life is moving on. It gives us satisfaction that we may have played a small part in enabling this to come about.

References

Dominica, F. (1982). Helen House—a hospice for children. *Maternal and Child Health*, 7, 335–9.

Dominica, F. (1987*a*). The role of the hospice for the dying child. *British Journal of Hospital Medicine*, **38**, 334–43.

Dominica, F. (1987*b*). Reflections on death in childhood. *British Medical Journal*, **294**, 108–10.

8 *Martin House*

LENORE HILL

The Martin House appeal was launched in February 1984 by some local paediatricians and the Vicar of Boston Spa in order to provide a 'Helen House' for the North. Built in Boston Spa near Wetherby, Martin House is only a mile from the A1 and is situated half way between Leeds and York.

Martin House has benefited greatly from the generosity of Helen House in sharing their experiences—not only of their successes but also of those improvements in facilities and accommodation which wisdom of hindsight and practice have made clear.

Our doors were opened to families on 14 August 1987 and by the end of the first year of being fully operational, Martin House had gathered the following referral statistics (Table 8.1–8.3). These figures include some children who were known to us before we opened.

Table 8.1 Referral statistics (14.8.87–13.8.88)

Number of children referred	108
Age range	3 months to 24 years
Children referred but not accepted	6
Children who died at Martin House	10
Children who died elsewhere	14

These are cold figures, saying nothing of the distress and heartbreak for sick children, parents, brothers and sisters, and other family members and friends. They are trying to live through what must be one of the worst experiences any family can be asked to endure.

Out of the 102 accepted referrals, the number of children who have stayed at Martin House is approximately 78. All the others received support and help in their own homes (as have also the families of the children who have stayed with us). The

Table 8.2   Categories of disease (August 1988)

Metabolic disease	27
Neoplastic disease (CNS 10)	15
CNS degenerative disease	15
CNS other	11
Progressive neuromuscular disease	15
Congenital heart disease	9
Chromosomal abnormality	6
Cystic fibrosis	3
Other	7
Aplastic anaemia	1
Post viral syndrome	1
Fibrosing alveolitis	1
SLE and multiple sclerosis	1
Bronchopulmonary dysplasia	1
Ondines curse	1

Table 8.3   Families and children using Martin House—(14.8.87–13.8.88)

Families with more than one affected child	8
Children currently using Martin House	47
Bereaved families currently, actively using Martin House	19

NB 'Currently' refers to August 1988

involvement at home varies from telephone calls, cards, letters, and occasional visits, to daily visits, a member of the team actually staying for a few days at the home of a family, or to helping to care for a dying child. We also try to visit if a child is in hospital and have had staff resident with parents in hospital at particularly stressful times when a family has requested this.

Our aim is to be regarded by our families as friends rather than simply as 'professionals' (Table 8.4). We have been amused and delighted to have once or twice been asked by tired mothers to send someone from our care team to help clean up—because 'the nurse' was coming and it would not do for 'the nurse' to find the house untidy! We also have volunteers working with

us and occasional student placements or some students doing
holiday jobs. Further staff are available to us through a 'bank'
system.

Table 8.4   Martin House Care Team: August 1988

RSCNs	12
RSNs	2
ENs	2
Physiotherapist	1
Social worker	1
Teachers	2
Nursery nurses	3
Psychology graduate	1
Chef	1
Unqualified	1

Volunteers, occasional student placements and student holiday jobs. Further
nursing staff available through a 'bank' system.

A valuable lesson we learnt from the experience of Helen
House was the sharing of household tasks. It feels more like
'family' if we are all doing the sorts of things which families
normally do at home. Parents can work out a lot of their own
stress in the kitchen—and the children love baking. I have had
some of the most profound conversations standing side by side
with a mum at the kitchen sink, or hanging out the laundry.
Sister Frances Dominica is often quoted as saying that hospice
is a philosophy, not a facility. The word 'hospice' is itself often
misunderstood. Like the word 'asylum', which began its life
as a beautiful word and became something ugly and unattractive,
so our word which means a place of hospitality—a place of rest
and refreshment—is being abused and spoiled. I have friends
in adult hospice work who also believe that they are involved
in quality of life, not simply providing a 'place to die'. Of course,
the word 'hospital' was at one time synonymous with 'a place
of death'.

Certainly the philosophy of 'hospice'—of real hospitality and
caring—should be available for all our families with sick children,
whether they have a life-threatening disorder or not.

One of our boys, in hospital within a few weeks of his death, and physically unable to do anything for himself, was scolded for pressing his nurse-call bell for help. He was also told when he expressed to a nurse in the middle of the night his fear that he might die, that he would 'have to get used to that idea'.

I know that there are many caring staff working hard in hospital wards, but we should not tolerate these attitudes in anyone working with the sick and their families.

Facilities are also important. I spoke to the uncle of a child who had just died at Helen House on my first visit there. He told me what a difference it made to all the family that they had been able to be comfortable and not sitting on hard chairs in draughty hospital corridors. Let us continue to campaign for adequate facilities for all relatives staying with a sick family in hospital—comfortable beds, adequate washing facilities, meals on the wards, provision to make a drink whenever they wish to do so—the comfort and privacy most of us take for granted in our day to day lives.

Of course, the facilities we provide in a childrens' hospice are purposely very different from hospital—our function is totally different. Children and parents come to us to enjoy a break away from their own four walls: to enjoy a holiday—a homely environment which they can use together as a family. One of the many pieces of invaluable advice which we had from Helen House was that we needed plenty of accommodation for parents and siblings (uncles, aunts, and grandparents too), when we have a child with us for terminal care.

For a large number of our families, the facilities which Martin House offers are accepted because they are available to all members of the family. Indeed, we believe that we function at our best in caring for whole families. At times, *all* family members *need* to be in a relaxed and caring environment in order to cope. With us they can know that their sick child is safe—being cared for, not by 'experts' but by 'friends' who will carry out care in the way they have instructed. As parents learn to trust us they will begin to go out sight-seeing, shopping, or for a meal together. The novelty of a night of undisturbed sleep can be followed by a day with no worries about dirty clothes, vacuum cleaners, or what to have for lunch. There are of course, some youngsters who enjoy a break away from their families. They enjoy the feeling of independence, and they know that they have us under much better control than they have their parents! Some parents too are looking for some space to be alone together, or for more time with the well brothers and sisters.

Watching, as we do, the slow deterioration of some of our children we have approached and seen into the chasm of the parents' sufferings. They invariably bring pictures with them of their children when they were well, and want to share with us their children's achievements and the things they used to do.

We can certainly all identify with the old saying 'We were given one mouth and two ears for a reason'! We need to listen and to help parents to listen to each other and to be able to break down some of their mutual protection barriers. We can never say, however, as our families can say to each other, 'I know how that feels, I've been there'. Our parents are an enormous support to each other.

One of the mothers said one day, 'Do you know what Martin House has done for me? It has helped me to find the person I used to be before this illness took me over'. Another mother said, 'To me, Martin House is a place to which if I'm tired or frightened, or David is ill, I can just run and someone will open their arms and welcome me in. I'll be loved and cared for; and no one will take David from me. I can go on loving him, and looking after him . . . if you will go on loving me'.

I know that the most controversial thing we do is care for children who are dying. 'The right place for a child to die is at home' we hear the 'experts' say. The right place for a child to die is where that child, or his parents, feel is the right place for them.

After her son had died with us, one mother said that she could never thank Martin House enough for being there. She knew that she would not have coped at home, and her son had hated hospital. She would have felt that she had let him down if he had died there. 'Fetching him here' she said, 'felt like fetching him to his other family.'

Andrew took his first look at Martin House and told his parents, 'This is where I want to die'. With a lot of support and help from the oncology unit where he had been treated, we tried to get him home—at least for a while. He withdrew totally, refused to get out of bed, or to speak to anyone. Once back at Martin House he had a full and busy life—even visiting Blackpool to see the lights. At his special request, he visited a motorway cafe for a beefburger and a cup of coffee, the evening before he finally lost consciousness.

Martin House is Christian in concept, and we are privileged to be working with the Sisters of the Order of the Holy Paraclete—an Anglican order whose mother house is at Whitby. We value greatly their daily prayers for our families and our work, and also the regular involvement of our chaplains. We have a chapel situated in the Sisters' House but accessible to anyone who wishes to use it. We are delighted, however, to have welcomed families to stay with us from Muslim, Hindu and Jewish faiths, agnostics, atheists, people who have never thought about faith at all—and even a few Christians!

To lose your own child must be the hardest thing for any parent to have to endure. If they can be left feeling that they did what was absolutely right for their child, then over the years to come it will help them (and their well children) to pick up the pieces and to move forward, building a new life for themselves; a life in which the life of the child they lost will always be a part. The friendship and support of Martin House will continue to be available to our families as they build that new future, until they eventually grow away from us, beyond their need for us.

9   *Acorns*
SALLY DAY

For some years in the West Midlands, there had been a feeling amongst professionals that there was a great need for respite care for children with a limited life expectancy. In late 1983 a series of events, with a catalyst in the form of a comment made by the Bishop of Birmingham, led to an open meeting being organized under the auspices of the Calthorpe Association for the Handicapped. This charitable association worked to support and develop the community aims of the Calthorpe Special School Youth and Adult Centre. The teaching staff at the school were responsible for providing much of the help and support needed by the parents of handicapped children in that school, and whilst they were very happy to do this, they felt that there was a need for a more permanent and extra service.

As a result of that open meeting, a working party was set up with a brief to identify and quantify the need for care and to recommend a suitable scheme to satisfy at least some of the requirements. The working party reported back in early 1984 and it was agreed that the need had been identified sufficiently that a scheme, providing a children's hospice for respite as well as terminal care and a home support service should be developed for the West Midlands. Trustees with a range of appropriate skills and backgrounds were identified, and the Trust was set us as a registered charity named the Children's Hospice Trust. Acorns, the children's respite hospice based in South Birmingham was established by the Trust and opened in 1988.

Although need had been identified, it was still difficult to plan this type of service since statistics collected by Health Authorities do not provide sufficient information. The working party had to rely on questionnaires sent out to paediatricians in the West Midlands and on views collected from staff in schools for children with special needs. They believed that within the defined catchment area of a radius of 25 miles based on the centre of Birmingham there would be between 400 and 500 families who may have need of this service. We are now planning to cover the geographical area of the West Midlands Regional Health Authority—comprising 22 District Health Authorities – and we believe that in that area there are likely to be between five and six hundred families who may wish to use our service. This was not a very scientific evaluation, but was corroborated by data published in *Care of dying children and their families* (Thornes 1988).

A major influence on services in the West Midlands is the variety in culture and religion between different ethnic groups. Many of the most needy families in the eyes of the providers of health care prefer to rely on their extended families and religious leaders for the care of their terminally ill young people. We hope to provide a service which these families will see as being relevant, existing alongside them in their traditional ways of caring.

The service that the Trust wishes to provide is based on the needs of the families with life-threatened or life-limited children between birth and 19 years. I believe this to be a 24-hour availability of a caring person who knows the child—or who at least has a good knowledge of the condition of the child and who has time to be a friend to the family. We aim to develop our Home Support Team to suit the family needs that are expressed as we begin to gather practical experience; but we start from the premise that we shall need a number of RSCN health visitors, other professionally qualified personnel with helpers or volunteers who are not professionally qualified. At present we have divided the region into four geographical areas and we have one staff health visitor operating in each area with one social worker visiting the entire region.

To provide a community-based home support service for up to 500 families in an area approximately 130 miles from north to south and 90 miles from east to west is a major challenge! All our plans will have to be extremely flexible until we can really see and understand the level of support being offered by statutory services, district by district across the region, and as we hear of the needs expressed by the families.

It is our intention to work co-operatively with all statutory services, and where necessary, be part of a 'package' of care provided to the families. With this in mind it is likely that a considerable portion of our Home Support Team's time will be spent in liaison with NHS, Social, and Education Services colleagues. We shall be monitoring very closely the proportion of time that individual team members spend in areas of direct family contact, travel, and 'administration'. In a year's time we shall be in a far better position to plan the future.

The Team members are now busy developing contacts with paediatric and community staff and social workers throughout the region. Happily they are all being well received by families and professionals alike, and are invited to be involved in several aspects of care, from attendance at case conferences, addressing meetings and seminars, meeting with specialist teams, and accepting individual referrals. It says much for the time and effort put in by the Trustees over the years that we have a great deal of support for what we are now doing.

The Home Support Team members are the first point of contact for families and referrers. We accept referrals from any source, but always seek agreement to our involvement with individual families from their general practitioner or the hospital consultant most involved with the child's care. At the same time the local general practitioner on call to Acorns liaises with his medical colleagues.

On the basis of initial information received on referral, a decision is made whether or not to visit the family at home for referral assessment. Once such a visit is made the decision to accept the child and family into the service is made by a committee comprising the head nurse, home support manager, general practitioner, social worker and hospice director, or in the event of an urgent decision needing to be made, by any two of those people. The care given to families will be in accordance with their wishes, and will extend to bereavement visiting for as long as appropriate.

Staffing

The Home Support Team leaders have already been discussed, but in addition to these four health visitors I anticipate employing district nurses, social workers, and others to enable us to fulfil families' needs with the most appropriate expertise. To introduce unqualified people or volunteers into families' homes has its challenges regarding compatibility and insurance for instance. Nurses and health visitors, by belonging to their professional organization, have their own professional indemnity insurance. Other staff members have no such cover and to provide professional indemnity insurance to protect staff working away from the Acorns premises is an additional cost. We shall overcome this by encouraging families, where appropriate, to make formal agreements with one or two workers so that they, the family, are taking the responsibility and liability for those workers. This has been found to work well by another voluntary group operating in the city.

Within Acorns, we have a multi-disciplinary team comprising nurses who hold the RSCN qualification, nursery nurses, a cook, a housekeeper, a general practitioner part-time, and other unqualified staff. Our social worker is based at Acorns but has a role related to the families both within Acorns and in their own homes, as well as a role with staff. The team concept is a rather difficult one for people used to a hierarchical management structure to come to terms with, but progress is being made, and individuals are beginning to be less protective of their specific qualification or training.

Education and research

It is one of the stated aims of the Trustees that the Acorns Care Team should develop an expertise and data bases which can be shared with colleagues. In order to achieve this aim we have been very careful to develop structured systems for all record keeping, and to do this in such a way that our work can be properly evaluated in the future. We also look forward to being able to host seminars, workshops, etc. to draw together for mutual benefit all the expertise we know to be available in the West Midlands.

Systems technology

We made an early decision that we need to be efficient and cost effective in all areas of administration, and that we should depend on computerized technology as far as possible. We have had the benefit of a computer consultant who is totally committed to the aims of the Trust and he has developed systems tailor-made to our needs. We believe these systems will have a commercial appeal to other small health care organizations. If this is the case our overall financial return should more than cover the cost of setting up.

In part, the importance of computer technology makes our organization sound rather structured and technical in its approach to the issue of providing hospice care to families whose children have a limited life expectancy. I would like to emphasize that we feel that a well structured system enables the staff to provide the care in a relaxed way. We shall not at any time allow computers, systems, or theories to rule us—they are merely tools to assist us in our stated aims of providing a caring, friendly, and relaxed service to families and children who have need of friendship from people who can empathize, sympathize but above all, help with the traumatic events which have overtaken their lives.

Reference

Thornes, R. (For the working party on the 'Care of dying children and their families') (1987). *Care of dying children and their families*. National Association of Health Authorities (NAHA), Birmingham.

10 *An evaluation of hospice care for children*
ALAN STEIN AND HELEN WOOLLEY

Introduction

When Helen House, the first children's hospice of its kind, was opened in Britain in 1982, it aroused considerable interest and it soon became apparent that a national children's hospice movement would develop. Since then, (as described earlier in this book) further hospices have been built. However, this has raised many questions and commentators have urged restraint concerning such developments (Chambers 1987) until important issues have been addressed, such as who uses children hospices and why.

Since the early 1970s, the adult hospice movement has increased in size as a result of the growing awareness of the needs of terminally ill people and their families. A recent comprehensive study in the USA of the work of the hospice movement (National Hospice Study 1986) has reviewed the efficacy of hospice care and whether or not it is 'cost effective' compared to more conventional hospital care for the terminally ill. In Britain, however, there has been relatively little evaluation of the work of hospices. The studies that have been carried out, for example those by Parkes (1979*a*, 1979*b*) and Hinton (1979) found that individual hospices were helpful in many ways to those people who used them. Moreover, the growth of the adult hospice movement has brought about other benefits. In particular, the caring professions have been made aware of the specific needs of people with terminal illnesses and much systematic research has been carried out into ways of providing symptom relief to people with intractable illnesses. Furthermore, this has also focused the minds of medical educators on the role of doctors in relation to terminally ill people. It has helped to change the focus of medical care, particularly for terminally ill patients, from 'saving lives' to improving the quality of life even if it is limited, from cure to palliation.

Despite these developments there has been little formal evaluation of the work of adult hospices in this country. We therefore considered it important to initiate an evaluation of the work of Helen House, from its inception, and were encouraged and enabled to do so by the staff. Our aim was two-fold: to assess the impact of chronic life-threatening illness on the family; and to examine the quality of care offered by the hospice, not only in objective terms, but also in terms of the family's perceptions. These two rather briefly stated aims included much else. It required us to characterize the families who used Helen House. Why did they come? Who referred them? What were their unmet needs? What care was available to them prior to coming to Helen House and how was it co-ordinated? How did they perceive the hospice and did they find it helpful? What specifically did they value about hospice care? We also needed to look at issues concerning the child's death.

The aim of this chapter is to describe some of our findings and bring together some of the issues raised by the research. While all the data has now been collected, not all of it has been fully analysed and some of our detailed findings are as yet preliminary. The study consisted of three component parts.

Retrospective study (Stein *et al.* 1989)

Firstly, we recruited a series of 25 families currently attending Helen House whose homes were within 50 miles of Oxford (a pragmatic decision, taken before the study commenced, to facilitate home visits to the families). Of the 25 families, 20 had surviving children (surviving group), and five had children who had already died (bereaved group). The bereaved group were not seen until a year had elapsed from the time of the child's death.

Prospective study

Two groups of matched children and their families were recruited for this part of the study. The Helen House sample consisted of 21 families consecutively referred whose homes spread across ten Regional Health Authorities. A control group was recruited from a number of sources, none of whom were receiving, or intended to request, referral for hospice care. The groups were individually matched for the child's diagnosis, age, and length of illness as well as social class. They were recruited largely from the Oxford Regional Health Authority Database of all admissions to hospitals in the region. Where it proved necessary we used other sources, such as the British Paediatric Association, Public Health Laboratory Service and the Society for Mucopolysaccharide Diseases registers. The control families' homes spread across eight different Regional Health Authorities. All families were seen oh two occasions. The Helen House families were seen at their initial 'look-see' visit to Helen House, in other words, before they had actually obtained any Helen House care. The second point of study was arranged for six months later. The matched controls were seen on two occasions, also six months apart. If a child had died between the two visits, the second visit was delayed until at least six months after the child's death.

There was a wide range of illnesses suffered by children referred to the hospice in our study. The most common illnesses among the children were: a group of neurodegenerative disorders about half of which had a confirmed diagnosis (for example Battens disease), the remaining half having no firm diagnosis; metabolic disorders especially the mucopolysaccharidoses, muscular dystrophy, and a small group suffering from neoplastic disease mainly of central nervous system origin.

Staff stress study (Woolley *et al.* 1989*a*)

Finally a study of staff stress levels and job satisfaction was undertaken. Twenty four of the twenty seven staff involved in the care of the children were interviewed to examine the level of stress experienced by the staff as well as to understand those factors which staff found stressful and those which helped them to manage in difficult circumstances. Levels of job satisfaction and related factors were also studied.

Results

At the outset an important question had to be addressed: is there something different about those families who are referred to Helen House when compared to control families? When we examined the data in our prospective study and the detailed transcripts of our interviews two broad trends emerged. There were a number of differences between the two groups, largely centring on the parents' perceptions of the quality of care they were receiving. However, the groups were also similar in many ways. They were particularly alike in respect to the marked impact of the childrens' illnesses on family life and the psycho-social functioning of its members. We shall deal with the differences and similarities in turn.

Differentiating features

Provision of services
The families were asked in some detail about what services were available to them as well as their views as to how helpful these services were in practice. We made a somewhat artificial division of the services into three broad categories: services provided in the community (such as general practitioners, community occupational therapists, physiotherapists, and social workers), services provided in hospitals, and services specifically for relief care.

Services provided by the community: Many families reported having access to a number of professionals. This varied greatly and included for example: general practitioners, community paediatricians, other medical specialists, social workers, community nurses, community paediatric nurses, health visitors, occupational therapists, and physiotherapists. However, in respect to the question of helpfulness, the Helen House referred group felt generally less helped by their general practitioners than did controls. The task of locating different sources of help from professionals and the various material benefits and provisions proved both complex and daunting for many parents of both groups. A particular problem for the families was how to combine the diverse help available while still feeling in control of their own lives with some choice left to them in managing the illness as well as living the time left together to the full. (Fig. 10.1). The problem was eased for two-thirds of the families who identified a 'cornerstone carer' or group of carers who acted

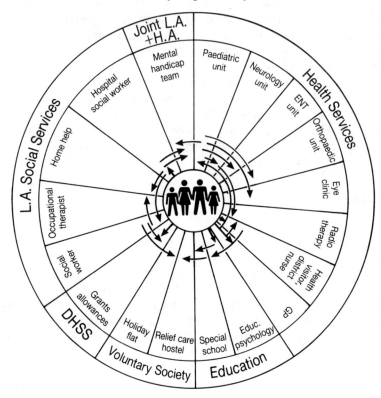

Fig. 10.1 The differing services encountered by one family an account of their child's degenerative condition. It notes referral sources for each service (arrows) and the administrative units under which the services come

as a central support and drew in other appropriate sources of help. The components of care especially appreciated by parents included: the personality of the carer enabling mutual respect; availability of predictable, mutually agreed contact that gave support without intruding; an ability to contain and live with the families' distress; an understanding of the problems associated with their particular child's illness; a working knowledge of available services (both personnel and material provision); the potential for continuity of care through illness and bereavement, and a sense of humour. Problems sometimes arose when the central care was invested in one person, because availability and continuity over a long period of time was not always possible. If the carer left without allowing for adequate hand-over time this could compound the family's sense of loss.

Provision of hospital services: Ill children from both groups were cared for by a wide range of specialists in geographically scattered units which, for 43 per cent entailed relating to more than one medical team, and for a minority (14 per cent) involved regularly travelling distances of 50 miles or more. With regard to the care received by the families at their local hospitals, both groups felt that they were receiving at least reasonable care. However, the control group felt significantly more supported and able to share their child's care with their local hospitals than did the hospice referred group. In particular the controls felt they were made more welcome at the local hospital treating their child and more involved with decisions about their child's care. Half the parents from both groups described a range of information which they would have liked: some forewarning about the course the illness might take, including the particular symptoms to be prepared for, and possibly some written information about the illness.

Non-hospice relief care: In trying to understand why families were referred to Helen House and wished to use the hospice, we felt it was very important to ascertain what sorts of relief care they had available, if any, and then to establish whether parents felt this care to be both suitable and sufficient. One of the most notable features was the wide range of relief care facilities used by the families and the vastly differing styles of the services on offer. These services ranged from a local authority hostel for mentally handicapped, a voluntary society hostel, over-night or boarding facilities at special schools, local or special hospitals, health service mental handicap units, a flexicare system where families were matched with another family offering relief care, night-nursing provision, and home care teams. A few families, however, wished to manage on their own with no outside respite care help. Some families also had supportive relatives and friends who helped with day and night time relief.

Whether the relief care on offer was acceptable depended upon a suitable matching of the style of service and the child's condition along with availability. The severely disabled child, for example with advanced muscular dystrophy, without intellectual impairment did not wish to stay in a local authority mental handicap hospital. A family whose child had had intensive hospital treatment for cancer was more likely to want to remain at home in the relatively symptom-free spells in between treatment and perhaps receive a home care team's treatment when problems recurred. Some special schools provided relief care as an extension of their education provision and shared the care of the child through that. There were however marked differences between the groups in the provision and adequacy of the care as well as their need for additional care if it was either not sufficient, not suitable, or not available at all. All of the hospice referred group wished for additional relief care. Six had relief care available to them but felt it was unsuitable, five had relief care available but felt it was insufficient, and the remaining ten reported that they had no relief care available at all. The control group presented a very different picture. Thirteen were using relief care and two felt it was insufficient, and another two felt it was unsuitable. This left nine who were receiving relief care who felt it was both suitable and sufficient and a further eight who were not receiving relief care but did not feel they needed it. A number of control parents, who at the time of the interview felt that the care available to them was both suitable and sufficient, did however express some concern over whether staff would be able to continue coping in the event of the child's further deterioration.

There were four main reasons why parents felt the relief care might be unsuitable. The first was their concern that the relief care facility was not geared to dealing with their child's symptoms such as pain, seizures, or feeding difficulties and that the relief care staff simply had not had sufficient experience of symptom relief in some of the very rare conditions from which these children suffered. Secondly, there seemed to be a 'mismatch' between the parents' wishes and the care offered, which inhibited the development of sufficient trust. Thirdly, parents sometimes felt that the relief care facility was too institutional and impersonal. Fourthly, many parents particularly wanted a facility where the whole family could stay. Where a child had been near to death once or more often, but had survived, it was hard for parents to leave their child in someone else's care overnight for fear of a sudden death in their absence. And yet many of these parents felt the need for a restorative break.

Our follow-up interviews, at least six months later, during which time the hospice referred group had had the opportunity to use Helen House, revealed some interesting changes. All but two of the twenty-one hospice referred families were satisfied with the relief care available to them. The first of these families felt that the hospice was simply unable to give them as much time as they wanted and the second lived a great distance away and also wished for some local overnight relief care. Interestingly, at the time of the second interview five families in the control group felt that they needed alternative respite care because the care available to them no longer seemed to be appropriate. Arrangements had broken down for four of five families participating in the family link scheme (whereby other families with well children cared for their children for brief periods during the year), and a further family felt that the respite care facility no longer had a grip on the child's symptoms.

The question of the quality and workability of the family link scheme is a very important one but is largely beyond the confines of our project. Nevertheless, a few points are worth making. First, the development of the family link scheme could potentially answer some of the essential needs of these families. The matching of the families is, however, quite crucial to the working of such a scheme. It is salutary that at follow-up interviews, four out of five control families, and one or two hospice families who were using such a scheme, had found the family to which they were linked to be incompatible. Clearly, it is impossible to extrapolate from the experience of seven families to the working of the whole scheme, but it certainly gives them cause for concern.

It is important to note that a group of control families were well satisfied with their relief care. The features of care most commonly appreciated were those of a welcoming home-like atmosphere in a non-institutionalized setting with experienced, committed staff, familiar with the child's symptoms and associated everyday problems: this allowed trust to develop and parents to feel secure in someone else continuing to care, albeit briefly, for their child. No family, however (control or hospice), had at the first interview a respite care facility where all the family could stay. Finally, it is important to recognize that a particular burden is placed upon families with more than one affected child, and support was provided for four of seven such affected families in the two groups via weekly/termly boarding school or unlimited time at a special residential unit.

Perceived adequacy of support and recent loss factors for families
At the time of referral, the hospice parents perceived the adequacy of their overall support to be significantly less than the control group. Eleven hospice families had lost the active support of a grandparent (usually a grandmother) in the preceding two years, either from death or serious illness or from withdrawal of contact. The actual news of the illness following diagnosis was so intolerable and painful to some relatives that they no longer wished to see the family. A small group of hospice referred mothers had recently suffered the onset of serious personal illness which compounded their already heavy burden of care. For those few who managed as single parents, coping alone with a dying child was indeed a very heavy load. Overall, it seemed that this loss of support and increased burden was an important stimulus for some families in seeking referral to Helen House.

Similarities between the groups in terms of impact of the illness on the family

Psychological distress
Both the mothers and fathers completed the general health questionnaire (GHQ)* in order to evaluate their psychological functioning; and the mothers filled in the Rutter-A2† questionnaires on their school-aged, intellectually intact children and their school-aged siblings. The notable feature of the results was the generally high level of psychological distress experienced by both

* The general health questionnaire is a 60-item self report measure assessing current psychological functioning. A score of 12 or above indicates significant psychological distress.
† The Rutter-A2 Scale is a questionnaire completed by parents about their children. It explores whether or not the child has emotional and behavioural difficulties. Examples of emotional difficulty are anxiety and unhappiness and examples of behavioural difficulties are temper tantrums and truancy.

the referred and control groups. This was most striking amongst the mothers, particularly of the hospice referred group, although there were no statistically significant differences between the two groups. Almost two-thirds of the hospice referred mothers were scoring above the cut-off point on the general health questionnaire. The fathers were also scoring relatively highly although this was not quite to the same extent as the mothers. Again, no statistically significant differences existed between the groups.

The general health questionnaire also contains a number of sub-groups—somatic, anxiety and insomnia, social dysfunction, and severe depression. Both mothers and fathers scored particularly highly on anxiety and insomnia as well as social dysfunction. Furthermore, 30 per cent of referred mothers and 14 per cent of control mothers were taking prescribed tranquillizers or antidepressants; and 38 per cent of the mothers and 14 per cent of fathers from both groups reported from moderate to great increases in their cigarette smoking habits.

The findings of the Rutter-A2 scores on the children's emotional and behavioural functioning suggests that both the intellectually intact children and their siblings were experiencing considerable psychological difficulty. Four out of six hospice referred ill children and seven out of nine ill control children were rated as having psychological difficulties on the Rutter scales. While these numbers are small, we feel it is important to note that these rates are at least double those found amongst children in the general population. Most of the difficulties experienced by these ill children were of an emotional nature such as anxiety and unhappiness. The siblings were also manifesting considerable psychological distress. Over half of the school-aged siblings from both groups were experiencing significant psychological difficulties. The major difference between these siblings and the ill children was that the siblings were experiencing both behavioural and emotional problems rather than emotional problems alone.

We also considered whether there was any relationship between the parents' and the childrens' psychological difficulties. In other words, were these difficulties clustering in the same families or in different ones. There was a clear relationship between the mothers' scores on the general health questionnaire and both her sick and well children's scores on the Rutter scales. However, there was little relationship between the fathers' and the children's scores or between the parents' scores. One explanation offers itself for these findings: the fathers, through their work outside the home, are perhaps more likely to maintain a psychological homeostasis and thus their GHQ scores are not well related to those of the rest of the family. There was nevertheless a relatively high level of psychological difficulties amongst certain fathers. This may be partly explained by the number of fathers who were unemployed or had given up work to help with the care of the child. Most of these fathers scored highly on the general health questionnaire while at the same time apparently providing an important protective mechanism for their wives' mental health. Thus a rather simplistic explanation may be that fathers who remain at home are able to help the family and support the family to a great extent, but end up manifesting much psychological distress themselves.

Impact on family's social routine and functioning
The detailed day and night practical care that some children needed set severe limitations on the family's ability to get out of the home together. Many parents had not had an evening out as a couple for months, and had very little time to get out with their well children. Exhaustion, worry over leaving the child safely in someone else's care, or finding a suitable minder made many parents, mothers in particular, feel very housebound. As one mother commented, 'I inhabit a circle from which there is no way out—my husband's gate is out to work each day. I am imprisoned in my own home and in hospital when my children are acutely ill'. One couple had not been able to get out of the house together for several years since their child's deterioration set in, and feared to leave her for more than a few minutes at a time in case she died in their absence.

We also made a more formal assessment of the families' social adjustment by administering the modified Social Adjustment Scale (Cooper *et al.* 1982). This enquires about each parent's social functioning in a number of areas: job, housework, social and leisure activities, extended families, children, and family unit. We found the families to have particular difficulty with respect to social and leisure activities in that they simply did not have enough time for these, and were having much less contact with their extended families than they might have had as well as having major difficulties in simply keeping up with ordinary household chores.

We enquired in detail about the functioning of the well siblings. At the time of the first interview two significant points of difference emerged when comparing the hospice referred siblings and the control siblings. A third of the siblings from the referred hospice group compared with no control siblings were reported as having fewer friends than their peers and more control siblings than their hospice counterparts were reported as having been introduced to the local hospital treating their ill brother or sister. Furthermore, a small group of siblings in both groups were described as having become excessively worried about their own health. This manifested in great anxiety about small ailments such as common colds.

Parental worry over child's symptoms
Parents were asked to rate the extent of their worries about a variety of possible symptoms which the children might be experiencing including pain, feeding difficulties, nausea and vomiting, breathlessness, swallowing, seizures, and drowsiness. Both groups of parents were worried about a wide variety of symptoms. The Helen House group were somewhat more concerned about symptoms (although this did not reach statistical significance) tending to be worried about pain, breathlessness, and appetite disturbance, while the control group worried about pain, breathlessness, and seizures. For some of the hospice referred group it was the acceleration of these symptoms which led to their seeking hospice referral. It was interesting to note that the apparent extent of symptomatology such as seizure frequency did not necessarily relate to the extent of the parents' worry. For example, many of the Helen House referred group had serious and sometimes intractable seizures yet only two parents reported great worry about their control.

We also asked the families about their perception of the ill child's problems with daily living. In particular, we asked about mobility, dressing, feeding, bedtime sleeping, toileting, communication and speech, play and social interaction, over-activity, and chronic irritability. Many parents reported multiple difficulties in these areas. When we went into the child's routine and the difficulties that the parents faced on a day-to-day basis, it gave us some idea of just what some of these families were facing. For example, some children needed major assistance with mobility, dressing, feeding, and toileting (being doubly incontinent), and experienced chronic irritability and difficulty in sleeping at night. There was one difference between the groups in that the hospice referred group were more likely to report greater difficulties with their child's mobility than did the control group. In fact for over half of both groups, mobility, toileting, communication, and speech posed major difficulties. A third of all children suffered persistent problems with bedtime sleeping. One parent with two children suffering from mucopolysaccharidosis could never get them to settle before midnight and would be disturbed up to a dozen times in the night before the day started at 6 am next morning.

Main parental worries
Parents were asked an open question about what their current main worries were. Although their responses were wide ranging, the most frequently expressed worry concerned the parents' anxiety about their capacity to continue coping with the child's illness particularly if their own health failed. This was most marked in the hospice group. The other issues were concern for the child's symptomatology and the course of the illness, and concerns about the child's ultimate death—where this would take place, what it would be like and how they would manage. A small number of families mentioned the impact of the illness on the lives of the well siblings although this was generally not regarded as a main worry for most of the families.

Financial burden and employment implications
A small number of fathers had been prevented from taking up a new job (e.g. promotion or transfer) because of their child's illness and overall seven fathers were unemployed, five of whom attributed this to their child's condition. Not infrequently fathers took time off work to help cope at home (26 weeks in one case) and some used up all their holiday entitlement in so doing. Two fathers' businesses went bankrupt shortly after the diagnosis, and they both attributed this to the financial outlay required

and time spent in caring for their child and the subsequent break-down in their own health. The mothers in both groups, however, had very high levels of unemployment. Several had given up their jobs after the diagnosis was made in order to cope with their seriously ill child. Of the few mothers who were in employment, most worked for no more than a few hours a week.

Families acknowledged that caring for an ill child particularly over a length of time could be very expensive. Typical expenses were for nappies (400 a month for one family with two affected children), travelling costs, heating, and laundry (£1000 a year for some families to keep their homes warm enough throughout the year), household alterations (extensions, ramps), and special equipment (hoists, stair lifts, wheelchairs). While the existence of a range of benefits (attendance, mobility, and invalid care allowances) grants and specific equipment had eased the load for many families, the difficulty frequently experienced in identifying and obtaining that help added to their distress. While half of the families had contact with a knowledgeable professional able to help them locate and mobilize available material help, others gleaned such information in a piecemeal way from voluntary societies or well informed friends. This left a further group still somewhat bemused by the need to negotiate help from so many diverse administrative sources. A plea for a central informant was made by many of this latter group. The financial and employment picture was similarly reported by families in the prospective and retrospective studies.

Impact of diagnosis
During the course of the interviews, one issue which came up over and over again was the parents' memories of the initial diagnosis and their preoccupations with the way it was conveyed. For many of the parents the diagnosis had been made many years previously, yet their memories of that interview lived on with a vivid immediacy, and they recalled minute details of the day they had been given the diagnosis. Just over half of the parents were satisfied with the way the diagnosis had been conveyed to them, but a significant minority were not. The parents' memories of the diagnosis interview and the importance they attached to the quality of that interview was not affected by

their satisfaction or otherwise. Many parents felt it was a key event which helped or hindered them in their subsequent acceptance, adjustment, and coping with the child's illness. Aspects of the interview which parents felt helped or hindered them subsequently, are reported in detail elsewhere (Woolley *et al.* 1989*b*).

In summary: parents particularly appreciated being given the diagnosis as soon as possible in an open direct way with sympathetic understanding. They wished to be seen together, in private, and uninterrupted, although some single parents welcomed the presence of a familiar professional (e.g. ward sister). Parents valued being given time so as to ease them through their initial shock reactions (anger, numbness, weeping), and found it helpful when the doctor repeated and clarified information in understandable language, making sure to ask parents what they had understood so far. Most parents wished for early information about the nature of the illness, its likely progression, and cause. How much parents felt able to absorb initially depended in part upon how long they had been worrying about their child's illness and whether they had suspected serious illness; the longer the prior worry the more able they seemed to hear and absorb the terminal news.

An outline of available support and help (a note of name, address, telephone number of helping services), an early follow-up appointment for further discussion and questions, and telephone number in case of emergency were reported as helpful.

Despite the deep shock of terminal news, parents wished to be fully informed about the illness, its likely progression, and the help available so that they could understand what they might have to face and plan for the future while knowing that they would be supported throughout the illness and death. Such knowledge and support lessened the parents' sense of helplessness and isolation and set up a therapeutic alliance for coping with the future.

Source of referrals to Helen House

While referral to the hospice of families from both retrospective and prospective groups came from a diverse range of professionals, self referrals formed the largest single group (just over a quarter) followed by paediatricians (one fifth) with only three of the 46 referrals coming from GPs.

Expectation for hospice care

Parents expressed a range of expectations for Helen House care. Although families had their own specific and individual reasons for seeking hospice care, a number of common themes emerged. Most common was the desire for quality care in a non-hospital or non-institutionalized environment with an emphasis on emotional support for the whole family and time to talk in a homely atmosphere: the quality of envisaged care being characterized by a combination of the staff's professional skills, personal care, and familiarity with the daily problems associated with their own child's illness.

The second need was for planned respite care, a facility flexible enough to enable shared care of their child with any or all family members who could stay or not as they wished. Where a child's condition had deteriorated beyond the scope of the local respite and support provision, parents hoped for somewhere offering pre-planned stays that were neither occasioned by nor cancelled because of an acute illness crisis. The added availability of an emergency bed at a phone call's notice with trusted staff in familiar surroundings was hoped for however to provide a safety net and lessen parents' sense of isolation. One family had never had a day's break—except for emergency hospital admissions—from caring for their 13-year old daughter who besides illness from birth also manifested severe behaviour problems. Another family whose teenager suffered from mucopolysaccharidosis had never been able to find a babysitter able to cope with his seizures, hyperactivity, and frequent irritability.

The third theme concerned the wish for medical care with an emphasis on the relief of symptoms rather than active intervention, particularly over pain, feeding, toileting, and seizures. A number of families also hoped to discuss and share the management of the problems with daily living posed by their child's illness, for instance, mobility and sleeplessness and to meet staff and other parents familiar with such difficulties.

A fourth theme concerned help with openly discussing and facing the future and in managing the child's death, whenever and wherever that might be. This included hoped for support for the whole family, particularly including well siblings in coping with the distress of witnessing the deterioration and death of one of their members.

Last, but not least, parents wanted their children to be at a place where enjoyment went hand-in-hand with continuity of the child's own preferred familiar routine and daily care. Most parents saw hospice support as augmenting any other services currently available and valued by them. Hopes for hospice care were summed up by one father.

'We need somewhere to carry on helping where other services have now left off: help with symptom relief, family support through contact or stays, respite care for our whole family to allow us to continue our care but share it. We want a hideaway breathing space and a place of beauty in which to die and a place of enjoyment where we can all make the most of life before death.'

Valued aspects of help received by families during the first 6–8 months of their contact with the hospice

Many parents' hopes and expectations for hospice care were fulfilled and often exceeded. In this section we discuss the valued aspects of hospice care most frequently mentioned by families: constant availability by phone and letters along with an available emergency bed provided a life-line for parents and lessened their sense of isolation; the freedom to talk openly and without fear about subjects which others so often avoided gave great relief; fears about what may be going to happen, how the illness may progress, about death, when, where, and how it may come, about practical issues around managing pain, seizures, feeding, toileting, sleep problems, coping with irritability or a constantly overactive and sleepless child, or an immobile child unable to communicate. The staff's familiarity with these issues through daily care of similarly afflicted children and their willingness to listen, to share and to talk, and to learn the family's familiar and daily routine of each child led to a growing sense of mutual trust. 'They took the fear out of death' was a frequently repeated comment of many parents and the fear of managing uncontrollable symptoms such as pain was eased as much by 'doing together' as by talking. Some parents were surprised and relieved at their reaction when someone else's child died while they were at the hospice. As well as easing their own fears, seeing other parents and siblings come through the experience, gave them hope for their own survival.

The home-like atmosphere in physically attractive and comfortable surroundings immediately put families at ease, and the willingness to share the child's care, as and when parents wished it, so that parents continued to feel in control was particularly appreciated. Sharing simple every day tasks and confirming and discovering preferred ways of looking after each child boosted parental confidence. With a child's lifespan so shortened, parents described a heightened sense of the quality of each day left to them: sharing that daily care, even allowing others briefly to undertake that care, was only possible if comfort and enjoyment was an integral part of high quality care. Parents appreciated the time spent in the 'ordinary' every day treats for their ill and well children: favourite games, shopping, outings, baking, model-making, listening to favourite music, and watching videos. Children greatly enjoyed these activities and many remembered them with great pleasure. Well brothers and sisters in one family referred to their ill brother's stays at the hospice as 'Will's treat'.

Flexibility both in domestic routine and in who could stay at the hospice was important. Some families took a 'holiday' altogether at the hospice, while others divided and went away for a holiday entrusting the care of their sick child to hospice staff. The natural inclusion of well brothers and sisters giving them time for enjoyment with staff, parents, and their ill sibling was valued. Parents described their pleasure in seeing well and ill brothers and sisters involved and enjoying themselves together, and this was an especially treasured memory once the child had died. This store of positive memories was important in helping sustain the family through subsequent grief: parents believed that this was especially so for the siblings.

Some simple, practical benefits were reported by parents: the knowledge of regular, planned respite breaks irrespective of whether the child was relatively well or acutely ill, enabled over half of the mothers to recharge their batteries and feel they could keep going at home in between times. Three mothers believed the stays had prevented their child going into institutional care.

Parents found they had time together, sometimes for the first time in years and some felt their partnership, so vital for coping, was strengthened. A third of the parents were able to take a holiday away for the first time with their well children and fewer fathers needed to use up their holiday entitlement in repeatedly staying home to ease through the difficult times. Three mothers felt that the break away from home had helped them to move out of a sickness-dominated world and to rediscover the more ordinary, forgotten everyday experiences again. A further three mothers described how previous constant worry, dread, and anxiety about the future had turned into a sense of positive planning for the next stage.

Nine Helen House families had lost a child and a tenth had lost two children during the six to eight months between our interviews. Six children died at the hospice, three in hospital and two at home. There were particular aspects of care around the death which drew favourable comment. The support for all the family before, at, and after the death without any sense of taking over was important; as well as help in involving any members, including extended family such as grandparents, cousins, aunts and uncles, important to the immediate family however near or distant the actual relationship, who wished to be there. At the hospice this included the siblings, many of whom visited and sat with their brother or sister in the cool small personal room set aside at the hospice for the family to be with and take leave of their child in whatever way they needed in the days between death and the funeral. Help with laying out their child in preferred clothes, with favourite toys and treasured possessions, with registering the death, making choices over the funeral, support over contacting distressed relatives and friends were all greatly valued. For most families, photographs taken around the death became particularly important in months to come and was a reminder of the quiet, natural slip into death. Attendance at the funeral of staff who had been closely involved with sharing the child's care was particularly welcomed, helping to carry the memory of the small everyday pleasures and moments of enjoyment for the ill child rather than only the disfiguring memory of illness. Letters, calls, and visits from Helen House staff when the funeral and business of arrangements was over and a sense of emptiness set in, was felt to have been very important. The opportunity to have time to talk events through, often repeatedly, and to share their grief with all the attendant fears was crucial. 'I can't concentrate', 'I can't think', 'I keep seeing him standing in the door', 'I hear his voice calling'. Such frequently described unfamiliar experiences along with periods of numbness, interspersed with periods of acute pain, led many parents secretly to fear that they were going mad. Parents needed to be reassured that, when they continued to grieve deeply beyond the first few months, there was nothing unusual about this, despite other people's expectations that they 'should have got over it by now'. Cards sent at anniversary times, such as Christmas, birthdays, and the first anniversary of the death were welcomed as a recognition of the continuing import-ance of the memory of the child.

Hospice care and perspectives

Our findings suggest that children's hospices can offer a special style of care to families of children with life-threatening illness. The hospice provides flexible care and support in a warm, friendly, unhurried environment, unburdened by the rapid turn-over and daily medical emergencies of the acute paediatric ward. This seems particularly appropriate for families under great strain, with children suffering from degenerative conditions for which curative treatment is not available or appropriate. The 'family-like' atmosphere with the option of 'whole family' invol-vement—especially important for siblings who are sometimes inadvertently excluded by conventional hospitals—is much valued by families. The hospice allows the whole family to be together instead of divided while focusing attention upon symp-tom relief and helping the family cope with their worry over symptoms. Particularly important here is the staff's familiarity with problems associated with the less common illnesses such as Battens disease or the mucopolysaccharidoses. Along with symptom control and management of disability, the emphasis is upon enhancing a child's enjoyment of whatever life is left to them: this means an individualized approach to each child and family in which the child's particular likes and dislikes are acknowledged, and their familiar preferred daily routine is continued. For many families the hospice is able to provide regular, pleasurable breaks free from the relentless 24-hour domestic routine.

Furthermore, families facing the isolating, long up-hill struggle of many degenerative conditions are able to draw support from other families while attending the hospice. In some cases, the hospice may have a special role around the time of the death, providing the whole family with privacy, continuity of support, and ample time in situations where dying at home is not desired or possible. Continuity of contact before, at, and following a child's death may help also to keep positive memories alive as the family grieves.

It should be recognized, however, that the actual nature of hospice care, in an environment with other terminally-ill children, could be a drawback for some families. Hospice care is in no way attempting to replace or detract from existing services; it is aiming to complement and augment them.

While there may be certain special features of care which are offered by the hospice, certain aspects of the style and nature of care could be further enhanced in paediatric wards. For example, ward staff could focus their care not only on the child but on the entire family, particularly the siblings; simply making them all feel welcome and encouraging them to visit regularly is greatly appreciated. In addition, attention to symptom relief needs to be directed not only to reducing the severity of symptoms, but to recognizing and allaying the parents' anxieties and helping *them* to cope with the specific symptomatology. Taking time to listen to the concerns of all the family (including siblings) is important, as is establishing a shared care ethos where staff and parents build up a mutual trust, and each mirror the best skills of the other in looking after the child, such that the family in no way feels any erosion of control over their own lives.

Furthermore, providing families with the opportunity to meet and draw support from each other may be helpful and while this may not always be possible on the ward, families may appreciate being put in touch with self-help groups, for example, the Muscular Dystrophy Group of Great Britain and Northern Ireland or the Society for Mucopolysaccharide Diseases (see Appendix). Families may also be appreciative of short episodes of respite care either in hospital or at a local respite care unit whichever seems more appropriate. It should be emphasized that many of these children face years of a deteriorating illness and that hospice, local respite care unit, and hospital are not mutually exclusive.

Our findings also indicate that a number of the problems, particularly the financial and employment ones go well beyond the remit of a hospice, and these problems are not easily resolved but at least warrant recognition. Other problems are more remediable via the adequate provision of local help, such as a capable relief carer to provide some day or evening care in the home, thus enabling the mothers and fathers to do essential shopping or keep appointments, or even allow a few hours of ordinary uninterrupted time with their well children. A flexible, welcoming local day care facility for a few hours or overnight stay may serve a similar purpose. Adequate night nursing provision may also give exhausted parents some regular uninterrupted nights sleep thus enabling them to keep going.

Finally, a diverse range of services from differing administrative sources, each with their own remit, may well be available to families over time in well serviced areas, but the problem of identifying and mobilizing that help can be considerable. The path can be smoothed here by a central, local, well-informed professional, or caring team able to act as service information source as well as helping to mobilize and liaise with those services as and when needed. Hospice care may indeed be one of those provisions drawn upon to complement a family's existing support. (see Fig. 10.1).

Staff stress and job satisfaction

As we point out in the introduction to this chapter, we felt it was necessary to take a broad view when trying to understand the functioning of the children's hospice. When we were planning the study, it soon became clear, that the psychological well-being of the staff was central to the functioning of the hospice. Thus our aims in this part of the study were: first, to examine the degree of stress experienced by staff caring for the children and their families; second, to enquire about those factors which staff found stressful, and those which helped them to manage in difficult circumstances; third, to gauge levels of job satisfaction and associated factors.

Essentially we found that the levels of stress among the staff were comparatively low as measured by psychological dysfunction and rates of absenteeism. Three-quarters of the staff were experiencing comparatively little stress and, in general, showed very few psychological symptoms, but a distinct sub-group were clearly manifesting a great deal of stress. A number of factors, notably recent personal bereavement and unresolved grief about a death that had occurred before they came to work at the hospice, distinguished this small group. Job satisfaction was generally high.

The main sources of stress were: the sense of impotence staff felt when they were unable to relieve perceived needs or distress; dealing with negative responses in families, and conflicts within the staff group. The most important mitigating factors were: the informal support which staff provided for each other in this small cohesive working unit, the home-like atmosphere of the hospice and the diversity of professional and personal skills among the staff group. The informal support which the staff provided for each other was mentioned universally by all the staff as the outstanding factor in helping them to cope through difficult times. They also emphasized the importance of feeling well supported by family and friends, and of being given (and allowing themselves) enough time to pursue outside interests, thereby maintaining a balanced perspective at work. Time and space for thought and discussion helped staff to avoid unhelpful over-involvement and allowed them to come to terms with the distressing events and situations frequently witnessed as part of their everyday work.

Although the total number of subjects was comparatively small, the findings have some general implications for both staff selection and staff support. In particular, when selecting staff, it is important to be alert to the possible vulnerability of potential staff carrying unresolved grief, as great distress can be rekindled when a current trigger event echoes back to and resurrects a sense of loss. Specific consideration and support needs to be given to staff who experience personal bereavement while working at a hospice. The very nature of the work serves as a constant reminder of their loss and many interfere with their own natural grieving. In particular, recognition of their grief, support, and time off may be necessary.

We believe that these findings have implications for groups of staff caring for terminally ill children, including those on paediatric wards, neonatal, and intensive care units. Together with the recent upsurge of interest in children's hospices, however, there has been a concurrent movement towards caring for the dying child at home. Should this trend develop further, the issues raised by this study will need to be tackled in the community (see section Personal reflections and domiciliary care pp. 9–46) For instance, a community paediatric nurse would need a base where he or she could draw support and understanding to help cope with the stress inherent in their work. It may well turn out to be that, while some families may prefer their children to be cared for at home, it could be much more difficult for the nurse caring for them. We believe that these are issues which will need to be considered by researchers and educators in this very difficult area, especially those concerned with paediatric nurse training.

In conclusion, differing styles of service are evolving for terminally ill children and their families to cater for a wide range of illnesses and associated needs. Although our reflections are specific to our findings from the Helen House studies, we hope they may serve as reference data in counselling service developments and particularly the need for further research.

References

Chambers, T. L. (1987). Editorial. *British Medical Journal*, **294**, 1309–10.

Hinton, J. (1979). Comparison of places and policies for terminal care. *The Lancet*, **8106**, 29–32.

National Hospice Study (1986). *The Journal of Chronic Disease*, **39**, 1–61.

Parkes, C. M. (1979a). Terminal care: evaluation of an inpatient service at St Christopher's Hospice. Part I. Views of surviving spouse on effects of the service on the patient. *Postgraduate Medical Journal*, **55**, 517–22.

Parkes, C. M. (1979b).Terminal care: evaluation of an inpatient service at St Christopher's Hospice. Part II. Self-assessment of effects of the service on surviving spouse. *Postgraduate Medical Journal*, **55**, 523–7.

Stein, A., Forrest, G. C., Woolley, H. and Baum, J. D. (1989). Life threatening illness and hospice care. *Archives of Disease in Childhood*, **64**, 697–702.

Woolley, H., Stein, A., Forrest, G. C. and Baum, J. D. (1989a). Staff stress and job satisfaction at a children's hospice. *Archives of Disease in Childhood*, **64**, 114–18.

Woolley, H., Stein, A., Forrest, G. C. and Baum, J. D. (1989b). Imparting the diagnosis of life threatening illness in children. *British Medical Journal*. **298**, 1623–26.

Acknowledgements

The study upon which this chapter is based was carried out in collaboration with Dr G. C. Forrest and Professor J. D. Baum. We wish to thank all the families who generously participated in these studies, the staff of Helen House for their co-operation, Mrs Ann Day for help with computing, Mrs Carolyn Fordham Walker and Miss Angela Sheppard who helped prepare the scripts. The study was funded primarily by the DHSS as well as by grants from The King Edward Hospital Fund and Help the Hospices.

Part IV

Specialist organ/system failure and parent self-help groups

11   Children with life-threatening congenital heart disease

HYAM JOFFE and HELEN VEGODA

The rapid advances in investigative and surgical techniques during the last two or three decades have resulted in the survival of many children born with congenital heart defects who, previously, would have succumbed in the neonatal period (first month) or in early infancy. Many of these children have complex cardiac anomalies and may need several high risk operations. In some cases, the condition can be partially treated but not cured. As a result, there is now a new population of children who face more than one life-threatening procedure and who may survive only until adolescence or early adulthood.

This chapter assesses the nature of the emotional support required by these families, the steps taken to address this growing need, and the magnitude of the problem in Bristol which is one of nine supraregional centres for infant cardiology and cardiac surgery in this country and covers the South Western Region, the northern part of Wessex and South Wales.

Background

About one of every one hundred and twenty babies (8 per 1000) is born with a congenital heart anomaly. Half of these are minor defects which do not cause symptoms or interfere with normal growth and development; these conditions do not require surgical intervention. Examples include mild obstruction of the outlet valves of the heart (aortic or pulmonary stenosis) and small communications between the two pumping chambers of the heart (ventricular septal defect). A number of asymptomatic patients will need corrective surgery during childhood to prevent the adverse long term effects which will arise if the defect is left untreated; such conditions include narrowing of the major artery to the body (coarctation of the aorta) and a communication between the upper chambers of the heart (atrial septal defect). The results of surgery in these patients are generally very good.

About 3 in every 1000 babies are born with complex lesions, resulting in severe symptoms soon after birth, such as extreme cyanosis (blueness) from obstruction of blood flow into the lungs (pulmonary valve atresia) or the passage of unoxygenated (blue) blood from the right side of the heart to the left (atrio-ventricular septal defect); there may be marked breathlessness from pulmonary congestion resulting from impaired blood flow to the body (extreme aortic stenosis). It is this group with life-threatening disease which imposes the greatest stress on patients, parents and families, and on the physicians and other staff involved.

The presentation of the problem often in the first days or weeks of life, is particularly distressing to the family. The symptoms develop rapidly and hospitalization is obligatory. The babies and parents are often transferred urgently by ambulance to Bristol which may be over a hundred miles from the family's home. The diagnosis in these critically ill babies can often be made by non-invasive, two dimensional echocardiography (ultrasound) so that immediate life saving surgery can be instituted. In some newborn babies, and in the pre-operative evaluation of older infants or children with complex disorders, more detailed information can only be obtained by invasive investigations such as cardiac catheterization, often performed under general anaesthesia. This requires the passage of a tube (or catheter) via a vein in the groin to the different chambers of the heart where pressures and oxygen levels are measured. Although the risks are low, parents often find this procedure particularly stressful since the future course of their child's life depends on the detailed findings of the investigation.

An operation in the neonatal period is often a temporary (or palliative) procedure to increase (aortopulmonary shunt) or decrease (pulmonary artery banding) blood flow into the lungs. Some conditions, such as coarctation of the aorta, can be corrected, or potentially cured, immediately. These procedures make up almost half of all heart operations in children and are called closed heart procedures since the surgery is performed on the major arteries outside the heart. Open heart surgery requires cardio-pulmonary by-pass whereby a heart-lung machine pumps blood around the body while the surgeon operates on a bloodless, non-beating heart. Although many of these operations are curative, an increasing number of complex conditions are tackled and these patients will require long term review, often for the rest of their lives, with the likelihood of further surgery in future years. Patients are reassessed at regular intervals by the cardiologists, either in Bristol or in one of the sixteen peripheral clinics scattered throughout the catchment area of the supraregional centre.

Although children with a congenital heart disease have a chronic disability, many are very well in the intervals between investigative and surgical procedures and, indeed, may be able to lead normal lives throughout childhood. However, the spectre of high risk surgery, with operative mortality rates of up to 30 per cent in some cases, imposes a great strain on families in anticipation of the predicted crises.

During the last few years, it became evident that the volume and intensity of emotional support required by the rapidly increasing numbers of patients exceeded the capacity of the medical team, even when augmented by the extraordinary voluntary efforts of the Bristol and South West Children's Heart Circle (see Appendix). This resulted in a recommendation that a post of Counsellor in Paediatric Cardiology be created; this was enthusiastically supported by the Heart Circle who also gave their financial backing. The first incumbent commenced duties in January 1988.

The Paediatric Cardiac Counsellor's role

The Counsellor and Family Support worker makes contact with the 'heart families' at many stages along the road which may prove to cover many years, from initial diagnosis to corrective surgery. The difficulties faced by parents are varied and occur at different times. In this section our Counsellor examines some of the families' needs and her contact with them at these various stages.

Initial diagnosis and treatment

When families first learn that their child has a congenital heart defect, there are often feelings of shock, guilt, bewilderment and possibly denial that their seemingly perfect baby, carried through an often untroubled pregnancy, can have anything wrong. There may be anger or self-blame, with irrational reasons found for the defect, such as a quarrel or taking a single aspirin. For parents of newborn babies, especially a first child, there will be the disruption to the family plans and the loss of the longed for, perfect baby and all which that entails.

Parents may want to get away from the intimidating atmosphere of the hospital to the comfort of their own homes and surroundings and find their stay in Bristol, often many miles from home, an added stress. After these initial reactions, many parents can, and do, take in the information given more fully and begin to ask searching questions of nursing and medical staff. I introduce myself to the parents a day or so after admission; they usually want to talk about and understand the situation and think about practical arrangements for the immediate future.

Margaret and Douglas were devastated by the news that three day old Jamie, their first child, had a serious and complicated condition needing an immediate catheter procedure and possibly a shunt operation. They were too upset to take in who I was when I met them in the Intensive Care Unit, but the paternal grandparents seemed grateful for the support I was able to offer them as they cared for Margaret and Douglas. Jamie came through the operation well and I saw the parents briefly in the ward. They began coming to my room, talking of their fears that Jamie might have died. Margaret admitted her anxieties about taking Jamie home as she felt nervous of handling him and was unsure whether she would cope. She eagerly accepted my offer to contact the health visitor and to remain in touch with her over the months.

Intervening period

Most families take their child home days or weeks after the heart condition is discovered, with arrangements for regular outpatient check-ups locally by the paediatrician and cardiologists and the prospect of 'wait-and-see' before further treatment plans are made. This intervening period can be very stressful. Most families try to lead positive lives incorporating the heart child into the family group, especially where the child is active or developing normally. For others, arrangements and activities become centred around the heart child; holidays and outings cannot be planned in advance because of uncertain or ill health and worries about admissions to hospital. Siblings may resent the attention given to the 'special child' and develop behavioural or emotional problems themselves.

Samantha, aged nine months, was a poor feeder, slept badly and developed frequent colds that resulted in her admission to the local hospital several times following her discharge from Bristol aged two weeks. Jillian, her mother, became increasingly anxious about her and unable to cope as a single parent with two other children, who began reacting badly to their mother's depression and preoccupation with Samantha. Jillian had always kept in touch with me by letter and telephone and seemed relieved when I arranged to talk to her health visitor and social worker to mobilize help locally. Samantha was later seen by the cardiologist who recommended that she be brought to Bristol so that a catheter test could be done.

The catheter

The admission for a routine cardiac catheterization, which is not itself a risky procedure is often a stressful period for parents as it may be the first time since the initial diagnosis that a detailed picture of the condition is given and—more important—how long it will be before corrective surgery is arranged. For those parents who have managed to put their child's health to the back of their minds and to get on with normal family life, the catheterization can suddenly bring back those first traumatic days and confront them with the reality of the heart condition.

Build up to an operation

With the catheter complete and a vague date set for surgery, parents usually resume their normal life; for many it then comes as a further jolt to receive the letter of appointment to outpatients to discuss the operation with the surgeon. Up to this point parents have usually only met the cardiologist with whom they develop a strong bond of trust and confidence. They now have to meet the surgeon, in whose hands they must place their child, and face the reality of surgery and the risks involved.

Helen's mother and father seemed well prepared for their meeting with the surgeon, having read widely over the four years since Helen's birth, arriving with a notebook full of questions. Despite their preparations, both were upset by this meeting, dwelling afterwards on how difficult they would find it to give their consent to the operation, even though they knew they had no choice if they were to give Helen the chance of a healthy life.

The operation

The time waiting for the letter to arrive announcing the date of admission can cause great tension for families, especially where a child is seen to be deteriorating, becoming bluer and more incapacitated. It is usually with a mixture of shock and relief that parents receive the letter and prepare to travel to Bristol for corrective surgery.

Families arrive on the surgical ward two to five days before the operation; during this time anxieties heighten and spirits fall as the reality of surgery hits them. To help families prepare for this admission, I offer a pre-admission tour of the ward where they have the opportunity to meet with the nursery nurses, play-leader, and other nursing staff who will care for their child. Despite this and the considerable preparations made, the day of surgery, their first entry into the Intensive Care Unit to see their child, and the recovery period afterwards, are inevitably times of great stress for the whole family.

Graham and Rosemary became increasingly irritable with each other as 14 month old Gemma lay in the Intensive Care Unit after surgery, often in an unstable condition. They had imagined that if she survived surgery, she would immediately be on the way to recovery. Fortunately, the close bond between them and their trust in the nursing staff, backed up by the emotional and practical support I could offer, helped them through each day.

Following surgery

Happily most children recover from surgery and go home to convalesce within a few weeks. However, some have complications that put considerable additional pressure on families.

Four year old Lesley had to remain in intensive care for several months and made slow progress. The family visited daily despite her mother's physical handicap and were delighted when Lesley began to come home for short periods, but were concerned about their ability to cope long term with Lesley's special needs. I organized a meeting between the parents and representatives of various community services, the outcome of which was the establishment of a network of supporting facilities to suit the family's needs.

Liaison with professional staff

My work involves close contact with support services in the community, liaising with nursing and medical staff in caring for families in hospital. Nurses are often very sensitive counsellors and are there at a time when vital medical information is given, to help parents understand and come to terms with what is said. I also act as a link between cardiologists and surgeons, particularly where parents are expressing anxieties that can be resolved by a further discussion with the specialist.

Liaison with voluntary supporters

Apart from the support given to families by the professional staff in hospital and the community, an impressive voluntary organization, the Bristol and South West Children's Heart Circle, has been supporting these families in the region for the past fifteen years. Jean Pratten, the founder, still visits families regularly in hospital and in their homes. She was instrumental in motivating the Heart Circle to establish and fund the post of Counsellor when it was recognized that the outstanding voluntary work of the Heart Circle needed to be supplemented by a full time worker. The Heart Circle not only gives support by helping families share their feelings with other families, but offers practical help and comfort through grants and transport to and from hospital. It has benefited all families by establishing and upgrading accommodation in hospital and in a well furnished hostel nearby.

Bereavement support

The majority of children do well but sadly, and perhaps inevitably, some die during or shortly after surgery. I am often with parents when a child dies, to offer practical as well as emotional help at this traumatic time. Bereavement support is an important aspect of the work and many parents are grateful for the link with the hospital where they faced the ordeal of the operation which, despite the risks, offered their child the chance of life, and appreciate the comfort offered by my continued involvement.

Brenda, aged 9 years, died in her parents' arms the day after surgery. I had formed a close bond with them during the short time I had known them and was with them shortly after her death, sharing the agonizing time waiting for the father's family to arrive to help them home. Over the next few weeks and months they telephoned frequently, wrote or occasionally came to see me, expressing their relief that they could talk about Brenda to someone who had known her and had been with them at the time of her death but wasn't family or a close friend. At their request, I arranged a meeting for them with the surgeon to help answer the questions that persistently nagged at them, and was there when they made that first painful return to the hospital without Brenda. A year on, I still communicate with them and remember Brenda's birth and death days, which I know they appreciate. They are now in touch with another bereaved family who were given their names and this has helped with their own feelings of isolation.

I am now hoping to establish a monthly bereavement group for families who can come to Bristol as I recognize that many wish to share their grief in a supportive atmosphere.

Conclusion

Children with heart defects differ from those with malignancies or progressive chronic disease, such as cystic fibrosis. Whereas the latter require long term, debilitating treatment, cardiac children may be reasonably well most of the time but have to face predicted and planned life-threatening crises. Hence the mode of care and support needs to take this into account, assistance being rendered at the specific times of crisis; however, the effectiveness of such intervention will be dependent on the relationship built up between carer and patient and family during the inter-crisis intervals. The Paediatric Cardiac Counsellor, the first post in the UK designated as such, has proved of value in providing this care to the benefit of patients, families, and clinical staff; the magnitude of this benefit has yet to be measured.

12 *When a baby dies*

GILLIAN C FORREST

Over the past few years, we have become increasingly concerned about the emotional impact on families of losing a baby in pregnancy or shortly afterwards. In 1988 there were over 8000 families bereaved in this way in the UK alone. Many vivid accounts of individual experiences are now available in books and papers and a number of careful follow-up studies of such families have made it clear that the grief following the loss of a baby is similar to that which follows the loss of any loved person (Giles 1970; Wolff *et al.* 1970). There is an initial period of numbness, shock, and disbelief which may last a few hours or days, followed by intense bouts of tearfulness, feelings of guilt, despair and anger (the 'pangs of grief'), searching for a cause of the baby's death, insomnia, anxiety, and social withdrawal which may last for months or years. Eventually most people reach a stage of resolution of their grief with a gradual return to emotional and social well being.

Some people experience more difficult grief reactions than this where they may be incapacitated for many years. One study suggests that this may affect as many as one in five bereaved parents (Nicol *et al.* 1986). These difficulties include prolonged grief, severe relationship problems with partners, existing children, or the next baby, or incapacitating anxiety symptoms such as panic attacks and agoraphobia. These atypical bereavement reactions have been reported more commonly among parents whose baby was stillborn. Some authors attribute this to the fact that the baby never lived outside the womb, so that the parents experience the same sort of difficulties mourning their baby as the relatives of people 'missing believed dead' in accidents and wartime (Lewis 1979). However, our own researches (Forrest *et al.* 1981) and information from other maternity units have confirmed that the mismanagement of a stillbirth itself can aggravate the 'unreality' of the event. The baby may be removed quickly without ever being seen by the parents; the hospital may make all the funeral arrangements; and the mother may be discharged home very rapidly without follow-up being arranged (Bourne 1968). Moreover, until recently the first name of a stillborn baby could not be officially registered, and there were anomalies in the benefits available to the mothers.

To try to avoid these most damaging bereavement reactions and help recovery from the psychological effects of perinatal death, various authors have made recommendations for the care of the families involved. Klaus and Kennell (1976), the National Stillbirth Study Group (1978), the Royal College of Obstetricians and Gynaecologists (1985) and the Royal College of Midwives (1985) have all produced guidelines in the last ten years to help the various professional staff involved provide the best possible management. Their recommendations share certain key features. Firstly, they emphasize the need to enhance the reality of the event by encouraging parents to see, hold, and name, and even photograph their baby. They also encourage the parents to be involved in the funeral arrangements for their baby, and to have a marked grave rather than placing the baby in an unmarked common grave. They all stress the importance of good communication between professionals and parents, with opportunities for discussions about the cause of death and post-mortem results, genetic and obstetric counselling as appropriate, and sympathetic listening by professionals while parents are given opportunities to express their grief.

The effectiveness of such care is clearly very difficult to evaluate. However, we carried out a study of support and counselling after perinatal death, modelled along the lines of these recommendations (Forrest *et al.* 1982). Fifty women took part in the study and were allocated either to a group which received routine hospital care, or to a group which received a planned programme of support and counselling. We followed up the women for fourteen months; after the first six months the supported group were showing significantly fewer psychological symptoms of grief than the routine care group. By fourteen months 80 per cent of all the women in the study were no longer suffering from psychological symptoms. This provides some support for the views of many parents who have valued informed and sympathetic care after their babies had died.

The rest of this chapter will be devoted to a more detailed account of the management of the loss of a baby in pregnancy or shortly after birth.

In hospital

Managing a stillbirth

The death of a baby in utero is very often first discovered or confirmed by a scan, and so it is essential that the technicians in the scanning room are sensitive to this situation and develop skills in adopting a compassionate and flexible approach. For example, they should be free to waive any rules to enable the mother to be accompanied by whomever she chooses. If the mother is to return home after hearing the results of her scan the clinic staff should ensure that she does not travel alone.

Many women are very fearful of the labour and delivery of their stillborn baby, and often have unpleasant fantasies about what is happening to the dead baby. One woman wondered how it would be possible to give birth to a dead baby at all; 'a decomposed shapeless lump of cells'. If they express these uncertainties and fears they are sometimes called 'morbid' or 'ghoulish' by staff. It is helpful to prepare parents for the delivery by giving accurate information about any abnormalities known to be present, including the skin changes to be expected. Parents should be encouraged to remain together so that they can help each other during labour. After the baby has been delivered the midwife or the obstetrician should suggest that the parents see and hold their baby. If any abnormalities are present these can be described first and then shown to the parents along with all the baby's normal parts. (In our study, when describing this six and fourteen months later, the parents of deformed babies all focused on their normally formed parts and none had found the experience of seeing their deformed baby horrifying or distressing.) If the parents do not wish to see the baby at all, a photograph should be taken and kept in the notes for possible use later. Many of the parents who decline to see their baby do, in fact, regret their decision weeks or even months later, and are then glad to have the photograph. They also should be encouraged to name their baby.

Afterwards many parents like to be left alone together for a while to share their grief in private.

If a mother has been heavily sedated during labour, or had a caesarean section, she needs an opportunity to see her dead baby later when she has recovered sufficiently. One of the midwives could bring her baby to her room (and the hospital management should allow this), or she and her partner could be taken to the room where the baby is laid out.

Managing a neonatal death

When a woman is in premature labour, or where there is known to be foetal distress, it is very helpful if the paediatrician can see the parents before delivery, so that he is a familiar face and can prepare them by describing the special care unit and its procedures. After the delivery the parents should be able to see their baby before he is taken to the special care unit even though this may have to be very rushed because of the baby's poor condition. An early photograph of the baby can be a great comfort for the mother to keep at her bedside and for the father to have at home. Both parents should be encouraged to visit the unit as soon as possible to see and touch their baby; they value being involved in routine care of the their baby like nappy changing or tube feeding. The staff need to make every effort to keep them closely informed of the baby's progress.

When a baby is critically ill and there is no hope of recovery, the news of the fatality or the baby's condition should, whenever possible, be broken to the parents in private so that they can release feelings without inhibition. They will initially experience shock and numbness, followed sooner or later by intense feelings of grief and protest. At this time, or later, the parents will have many questions to ask about what has gone wrong and the information may have to be repeated many times over before they can grasp it. In our experience, parents in these circumstances are seeking explanations and expressions of sympathy from the staff, and a chance to ventilate their own feelings.

While the baby's condition is steadily deteriorating, parents may begin mourning before he has actually died and may find visiting him intolerably painful. In these circumstances, they will need a lot of support and understanding to maintain some contact. This situation may be difficult for the staff to accept but if their attitudes to the parents are critical, it will only increase the parents' anger, guilt, and despair and make it harder still for them to visit.

When the baby is known to be dying the parents should be told so that they can be with him. Some parents will want a minister of religion called in to baptize the baby, or perform other important rituals. If at all possible, the parents and baby should be placed in a separate single room with the minimum of equipment present. They can then be encouraged to hold him as he dies, and afterwards, if the parents wish, they can help with the laying out. One mother said to me, 'No baby should die alone. It was all the comfort I could give him, to hold him in my arms as he died'.

The post-mortem examination

The information provided by a careful post-mortem examination or necropsy, can be important in establishing the cause of the baby's death and in identifying congenital abnormalities and any genetic implications. However, many parents distraught at losing their baby, find consent for this examination very difficult. One mother put it like this, 'She's been through enough. Must she be cut up as well?' Parents can be helped to overcome this instinctive reluctance to consent by having a doctor explain to them that the examination is necessary to clarify what has gone wrong and if it is likely to happen again. It follows from this discussion that the results of the post-mortem examination will be very important to parents, and it is essential that they are given an opportunity later on to hear the post-mortem findings.

After the death or stillbirth

The parents will often be in a state of shock or numbness for a few hours or days before they can accept the reality of their baby's death. One woman, describing her reaction to her still-born baby said, 'We looked at her in our arms and I thought, she's asleep. Why doesn't she just wake up now?' Another said, 'His mouth came open as we held him and I thought, he's coming alive!'

Although it is important for medical staff to give explanations to the parents at this stage of what seems to have gone wrong, it is likely that these will have to be repeated later, perhaps several times. It may not be until the follow up appointment several weeks later that parents are able to take in the information.

Wherever possible the mother should be given the choice of a single room, or to return to her own ward. It is perhaps surprising that some women prefer to return to their own wards; it seems that the support of the friends they have made there outweighs the painfulness of close contact with new babies. A few certainly wish to have contact with other babies straight away. 'I've got to face up to it, and I'd rather do it now than postpone it.'

Some parents like to have information written down and the leaflet, *Loss of your baby* (National Stillborn Study Group 1978) is useful. Parents should also be offered the telephone number of the Stillbirth and Neonatal Death Society (SANDS) a voluntary society offering support to other parents (see Appendix) and details of any local administrative arrangements. The midwives should try to help parents express their feelings, and not retreat if they come upon the mother (or father) in tears. Lactation must be discussed; many women wrongly assume that they will not lactate if the baby has died. Help should be offered with the necessary registration and funeral arrangements (see below).

Discharge should not be hurried. Contact should be arranged with the hospital social worker or other designated member of staff, who can offer support while mourning is being established and assess the couple's supportive network at home. The results of our study and the work of Raphael (1977) in Australia suggest that unsupported bereaved people are more at risk of atypical reactions, and justify the mobilization of extra support from, for example, clergy, health visitors, and the local SANDS member. The community midwife and the general practitioner should be informed of the baby's death as soon as possible so that they can arrange to visit the parents at home at the earliest possible opportunity.

Follow-up arrangements

These should be planned to ensure that the parents receive obstetric counselling, genetic counselling (where appropriate) and an opportunity to discuss the post-mortem results. These follow-up interviews are best arranged between three and six weeks after the baby's death or stillbirth—allowing time for the post-mortem results to be available, and for the parents to have recovered from the initial numbing impact of the baby's loss. The doctor can then use this interview to help the bereavement process by providing an opportunity for the parents to go over the events surrounding the baby's death and express their emotions about it (even when a post-mortem was not performed).

Registration and funeral arrangements

For most young couples this will be their first experience of bereavement and they are often bewildered by the complicated administrative procedures of registering the baby's death or stillbirth and arranging a funeral. We have found that by helping parents with these arrangements some of the distress can be relieved.

Registration

In the United Kingdom all live births, irrespective of gestation, have to be registered by the Registrar of Births, Marriages, and Deaths. All neonatal deaths (irrespective of gestation) and stillbirths (i.e. babies born dead after twenty-eight weeks' gestation) also have to be registered. Such babies then have to be properly buried or cremated. In some places, a branch of the Registrar's office is attached to the hospital, but in others parents have to travel to the central office in the town where the baby dies. They take the Medical Certificate of Cause of Death or Stillbirth, provided by the doctor, to the Registrar's office and there are supplied with the Certificate of Burial or Cremation which is required by the undertaker. Until 1983, it used not to be possible to register a first name for a stillborn baby, but efforts by the National Stillbirth Study Group, SANDS and others have succeeded in changing the regulations.

Funeral arrangements

The funeral can be arranged either by the hospital (with a con-
tracted firm of undertakers) or privately. The Department of
Health has directed hospital administrators to meet the cost
of any stillborn baby's funeral, unless the parents wish to pay
for this themselves; they cannot, however, meet the cost of a
funeral of a neonatal death. The exact details of the hospital-
arranged funeral vary up and down the country. In some areas
the babies are cremated; in others they are buried in a local
cemetery. In all cases though, the babies should have their own
coffin and a proper funeral held within a few days. The hospital
administrators should inform the parents of the time and place
of the funeral.

The advantages of the hospital-arranged funeral are that parents are relieved of the burden of making the arrangements,
and the costs are minimal. The disadvantages are that it denies
parents the chance of involvement in this important part of the
mourning rituals. In some cases parents are told that they cannot
attend the funeral (and this suggests to us that proper procedures
are not being followed). Sometimes they have great difficulty
locating the site of the grave later (if the baby has been buried
rather than cremated); they usually find that they are not allowed
to mark the grave. In addition, there are frequently many myths
and misconceptions about the hospital-arranged funeral, for
instance, that the babies are, in fact, disposed of in the hospital
incinerator; that they are buried in mass graves without coffins
or ceremony; or that the funeral may be delayed for weeks or
months.

To improve the care of bereaved parents it is clearly important
for staff to be able to give them accurate information about
the funeral arrangements that apply locally and discuss with
them their needs and wishes. In Oxford, a member of the administrative staff has been specially designated as the Bereaved
Welfare Officer to help parents cope with these administrative
issues.

Sometimes—and increasingly—parents of babies who are
born dead under twenty-eight weeks' gestation, ask if they can
hold a funeral. There is, in fact nothing to prevent them from
doing so. In place of the Certificate of Burial or Cremation,

the undertaker will require a note from the attending doctor stating the gestation of the baby, the (presumed) causes of death, and authorizing the undertaker to proceed with the burial or cremation. We feel it is important to help parents with this if they feel the need to hold a funeral for their baby.

At home

The course of mourning a new baby

Initially, grief is usually intense and constant, then gradually the 'pangs of grief' described earlier occur with lessening frequency. They tend to be precipitated by specific reminders of the dead baby, for instance, the funeral, the expected date of delivery if he was premature, the onset of the first period, the first Christmas or anniversary of his death; as well as by more general things—baby clothes, a piece of unfinished knitting, the baby counter in the chemist's shop.

Insomnia is very common but resolves spontaneously over the first few weeks. Ideas of guilt and self-blame may take over from the early searching for extraneous causes of the baby's death and intense guilt feelings may be experienced by women who have had a previous termination of pregnancy. (One woman was convinced that the angry jealous spirit of her terminated baby had taken revenge on this much loved and wanted baby. She eventually 'appeased' the spirit by buying a set of baby clothes for him and placing them in the coffin.)

Many couples experience difficulties relating to their friends and neighbours in the early weeks. They find themselves avoided by those who know that the baby died, and dread being asked about the 'new' baby by those who have not yet heard. Friends may not know that bereaved people need to rehearse the events around the death for a long time, and try to cheer the parents up by switching the conversation to a neutral topic. The parents then feel they are being 'boring' and may withdraw socially for many months until this need passes. Many bereaved mothers experience destructive feelings towards other babies for a long time; these are frightening and upsetting and may lead them to withdraw from any contact with children. They are normally reassured to find that these feelings occur very commonly, and pass with time.

Fathers tend to cope by plunging themselves into work activities as soon as possible. They appear to have shorter bereavement reactions than the mothers; in our study, 86 per cent had recovered from the psychological symptoms of bereavement by six months, compared with only 50 per cent of the mothers. For some couples this disparity in their grief reactions strains their relationship; for others it seems that the man readily takes on the supportive role, and they are drawn closer. Parents with other children at home often describe difficulties handling their questions about the baby, for example, a four year old asked, 'Why can't you go back to the hospital, Mummy and fetch her when she is not dead any more?', a three year old asked, 'What did I do? Hurt the baby so it went away?' The children's own grief reactions appear to be relatively brief, except when their parents—usually mother—remain severely depressed, lethargic, and withdrawn for several months. Then naughtiness at home or school, or withdrawal, sadness and preoccupation with death and coffins may be seen.

Parents need information and support about these various aspects of normal grief reactions to help them understand the process they are going through. Their general practitioner or health visitor is often best placed to fulfil this task.

Miscarriage and termination of pregnancy

Losing a baby in the early weeks of pregnancy is a very common event, but one that is still distressing, particularly if either there has been difficulty conceiving or a previous loss. Some women experience a full blown grief reaction after a miscarriage and need as much careful support and sympathy as someone whose baby had died later (Oakley *et al.* 1984). A significant proportion of women who have had a termination of pregnancy for social reasons or because of foetal abnormalities, suffer intense grief afterwards and not the sense of 'relief' that many people expect. Hospital staff and general practitioners can do much to help by acknowledging these feelings, and listening sympathetically to them. The Miscarriage Association and SATFA (Support

After Termination For Abnormality) can also offer support (see Appendix).

The next pregnancy

For many women recovery is marked by the safe delivery of another baby. However, anxieties have been expressed that women may seek another 'replacement' baby before they have sufficiently mourned the dead baby. It appears that pregnancy inhibits grief work and mourning, and so it seems sensible for women to wait long enough to mourn the last baby before conceiving again. In any case, they need to feel able to cope with the inevitable anxiety of the next pregnancy and delivery. SANDS suggests this may take nine months or more. It is likely to vary a great deal between individuals but the more strongly the identity of the dead baby has been established, the less likely it is that the 'replacement baby syndrome' (Cain and Cain 1964) will occur.

Conclusion

This chapter has outlined some of the problems associated with the loss of a baby in the perinatal period, and how medical and nursing staff can help parents to cope. More research is needed into the long term effects of preventing pathological reactions. The Perinatal Bereavement Research Unit at the Tavistock Clinic (see Appendix) has an important part to play in this area both in training and research. Medical care does not end with the death or stillbirth of the baby, and there is a great deal that can be done to alleviate parents' distress and facilitate their recovery from the loss.

References

Bourne, S. (1968). The psychological effects of stillbirth on women and their doctors. *Journal of the Royal College of General Practitioners*, **16**, 103–2.

Caine, A. C. and Caine, B. S. (1964). On replacing a child. *Journal of the American Academy of Child Psychiatry*, **3**, 443–55.

Forrest, G. C., Claridge, R., and Baum, J. D. (1981). The practical management of perinatal death. *British Medical Journal*, **282**, 31–2.

Forrest, G. C., Standish, E., and Baum, J. D. (1982). Support after perinatal death. *British Medical Journal*, **285**, 1475–9.

Giles, P. (1970). Reactions of women to perinatal death. *Australia and New Zealand Journal of Obstetrics and Gynaecology*, **10**, 207–10.

Klaus, M. H. and Kennell, J. H. (1976). *Maternal infant bonding*. C. V. Mosby, St. Louis.

Lewis, E. (1979). Mourning by the family after a stillborn or neonatal death. *Archives of diseases in childhood*, **43**, 303–6.

National Stillbirth Study Group (1978). *Loss of your baby*. Leaflet published in conjunction with MIND and SANDS by the Health Education Authority, London.

Nicol, M. T., Tompkins, J. R., Campbell, N. A., and Syme, G. J. (1986). Maternal grieving response after perinatal death. *Medical Journal Australia*, **144**, 287–9.

Oakley, M., McPherson, A., and Roberts, H. (1984). *Miscarriage*, Fontana, London.

Raphael, B. (1977). Preventive intervention with the recently bereaved. *Archives of General Psychiatry*, **34**, 1450–4.

Royal College of Midwives (1985). *Midwives and stillbirth*. Leaflet produced in conjunction with RCM and the Health Education Authority.

Royal College of Obstetricians and Gynaecologists (1985). *Report of the RCOG working party on the Management of Perinatal Deaths*.

Wolff, J. R., Nielson, P. E., and Schiller, P. (1970). The emotional reaction to a stillbirth. *American Journal of Obstetrics and Gynaecology*, **108**, 73–7.

13 *Children with AIDS*

JACQUELINE MOK

The first cases of paediatric acquired immune deficiency syndrome (AIDS) were described in 1982, in children of intravenous drug using and sexually promiscuous women. Although children can be infected through blood and blood products, the vast majority have been infected by vertical transmission. In the United States and Western Europe, the rise in paediatric AIDS cases parallels the rise in the number of infected women. Between 55–75 per cent of women have acquired the Human Immunodeficiency Virus (HIV) through needle sharing during intravenous drug use (IVDU), with about one third being heterosexual partners of infected men.

In the United Kingdom, the south east of Scotland has been shown to have a high prevalence of HIV infection amongst the IVDU population, one third of whom are women (Brettle *et al.* 1987). In Edinburgh, women at risk of HIV infection tended to be young and single, and to live in areas of the city characterized by multiple deprivation (Johnstone *et al.* 1988).

The Paediatric Counselling and Screening Clinic

A clinic was started in January 1986 at the City Hospital in Edinburgh to monitor the progress of all infants born to HIV infected mothers. It is held in conjunction with an adult screening and counselling clinic and it was hoped that all members of the family might attend. The life-styles of drug-using parents have meant that more and more home visits have had to be made and at present about 80 per cent of the children are seen at home.

The clinic is staffed by a consultant paediatrician, a health visitor, and a research registrar. One of the paediatricians usually meets with the woman in the antenatal period to advise on the risks of pregnancy to the mother and HIV transmission to the child. If the woman chooses to continue with the pregnancy, permission is obtained to see her child for follow-up.

One of the paediatricians is present at delivery to obtain cord blood and to examine the infant. Visits are then arranged either to the clinic or to the family home. The infants are monitored on a three-monthly basis and when required, facilities are available at the Infectious Diseases Unit of the Hospital, where admission is under the care of the consultant paediatrician. Through these procedures, continuity of care is maintained.

Diagnosis and clinical spectrum of HIV disease

Following infection with HIV, antibodies can be detected against viral core and envelope protein. The presence of passively acquired maternal antibodies, which persist into the second year of life, makes the HIV antibody test an unreliable marker of infection in a young child. Other tests, for example HIV culture, antigen or gene amplification techniques are still under evaluation. The diagnosis of HIV infection in children, therefore, rests largely on clinical evidence.

The initial presentation is usually with non-specific signs and symptoms; failure to thrive, recurrent infections, chronic diarrhoea, persistent candidiasis (thrush) and generalized lymphadenopathy (swollen glands). As the disease progresses, 50 per cent of infected children manifest neurological signs, others have lymphoid interstitial pneumonitis (a chronic pulmonary disorder thought to be a direct effect of HIV) while recurrent bacterial, viral, and opportunistic infections affect most children. The medical management of the infected child therefore varies with the clinical disease, but family and social factors often predominate medical issues.

Care of the infected child

In the absence of a specific cure, treatment is supportive and directed at the symptoms of HIV disease. Most infected children will live for years, although their survival is not without minor recurrent illnesses and repeated hospitalization. As yet, the natural history of paediatric HIV infection is not known, so that many families live with a disease of an uncertain future and a usually fatal outcome. Support must be ongoing, and able to bridge the gap between hospital and community services. Members of the multidisciplinary team should include physicians (for the child as well as parents), nurses, dietitians, physiotherapists, occupational therapists, speech therapists, social workers, and teachers.

Recurrent infections

Children with HIV infection have complex defects of their immune systems and are prone to repeated episodes of bacterial infection. Regular infusions of gammaglobulin (200–400/kg at 2–4 week intervals) have been shown to reduce infective episodes, increase weight gain as well as to result in decreased levels of HIV antigen in some children (Hague *et al.* in press). Other centres have used prophylactic antimicrobial therapy (specifically cotrimoxazole) and anecdotal reports appear encouraging. Controlled trials will be necessary to show the superiority of one treatment over the other.

Therapy with intravenous gammaglobulin involves regular attendance at hospital, albeit on an out-patient basis. Many of the children requiring this treatment come from families where lifestyle is chaotic, making it difficult for appointments to be kept. We have utilized voluntary agencies to provide transport for the families.

Candidiasis in the mouth or nappy area can persist despite topical therapy, and necessitate hospitalization for systemic administration of such antimicrobial agents as fluconazole, ketoconazole or amphotericin B. The most common opportunistic infection is *Pneumocystis carinii* pneumonia, which is usually treated with high-dose cotrimoxazole. Failure to respond, or the presence of toxic effects, means that treatment with other agents (such as pentadmidine) are required. If the lung infection progresses to cause respiratory failure the parents should be included in the decision as to whether assisted or artificial ventilation should be instituted. The outlook is generally poor, although some children have survived repeated episodes of *Pneumocystis carinii* pneumonia. Prophylaxis should be commenced following one episode of *Pneumocystis* pneumonia, using oral cotrimoxazole. Pentamidine delivered by aerosol may have a role and deserves evaluation in paediatric practice. However, compliance with prophylactic therapy—demanding medication provided by the parent(s) at home on a daily basis—remains in doubt.

Nutrition

Attention should be paid to the child's diet, even when symptoms are mild. Failure to thrive is probably a direct effect of HIV, and should be addressed early. Parents with limited incomes should be taught how to replace 'junk foods' with those of higher nutritional content. The child with candidiasis of the mouth will be reluctant to eat, and it may be necessary to resort to tube feeding, either continuously or overnight. With protracted vomiting or diarrhoea, total intravenous nutrition may be the only means of providing adequate fluids and calories.

With the help of the home care team, the child may be able to live at home despite indwelling catheters. In families where intravenous drug abuse persists, care must be exercised where supplies of needles and syringes are freely available.

Respiratory support

The pneumonitis of AIDS is slowly progressive, leaving the child with limited exercise tolerance, repeated chest infections and varying degrees of hypoxaemia. Regular monitoring of oxygen saturation will enable a decision to be reached on supplemental oxygen therapy. Again parents need supervision and support on the use of oxygen in the home. Severe limitation in respiratory function may require special education facilities for the child, ranging from withdrawal from sports at school, to the provision of a home teacher.

Brain involvement

Periodic assessments will usually reveal delays or regression in developmental milestones. Neurological involvement can progress rapidly or slowly, and be accompanied by seizures. The child may begin to suffer from both physical and mental handicap, and will require support from the physiotherapist, occupational therapist, speech therapist, and special educational staff. It is often necessary to liaise with staff from Education and Social Work departments, so that an optimum placement is achieved for the child with signs of HIV encephalopathy.

Terminal care

In a disease where therapeutic options are limited and the ultimate prognosis is poor, the parents do at times wonder whether the child is subjected to unnecessary suffering. When every treatment regimen has been tried and failed, parents may wish to take the dying child home. Unfortunately, a vast majority of the mothers will themselves be infected (see below), and their health may not permit them to nurse their child. It may be possible, with the establishment of supported accommodation units, for terminally ill mothers to be helped to look after their own children. Foster parents in Edinburgh have also expressed a willingness to nurse terminally ill children, with support from the home care team.

The family

There are few other instances where a potentially fatal illness affects both young parents and their children. HIV infection characteristically affects young men and women. Transmission of the virus heterosexually and from mother to child will result in whole families being infected. Paediatric AIDS has been termed a 'Family Unit Disease', and many of the families may be disorganized and functioning poorly prior to the diagnosis.

The majority of children have acquired HIV infection from the mother, who is likely to have been infected through IVDU. Continued drug use or ill health in the mother may result in alternative care being sought for the child. Other siblings may also be infected, and entire families comprising infected as well as non-infected siblings may require foster care as parents succumb to AIDS or continue to use drugs.

The extended family may be alienated because of drug use. Our experience is that many of the mothers had only shared needles on one or two occasions, unknown to their own parents. The secrecy surrounding the drug experimentation extends to the diagnosis of HIV infection, and leads to exclusion of support from grandparents. Many grandparents do rally around when they discover the diagnosis, and some have to cope with more than one of their children and several grandchildren being infected with HIV. Tragically, grandparents have had to bury their children and grandchildren; and they are often forgotten when support services are provided.

The stigma of HIV

AIDS was first described in the homosexual and IVDU population, which are socially disenfranchised groups. The general response has therefore been one of stigmatization and victimization of infected individuals, who have been refused mortgages and insurances, as well as lost employment and friends. Some families have had their houses burnt down, while children have been refused admission to schools and nurseries. Even healthy siblings have been denied school entrance and social contacts.

Parents therefore find themselves afraid to disclose the child's diagnosis with the attendant risk of social rejection. Many families withdraw in anticipation and refuse support when it is offered, in case their confidentiality is breached. As a result, staff are often asked to act in an advocacy role for the child when dealing with Social Work or Educational establishments.

Parents and foster parents are counselled to consider carefully before divulging the diagnosis to anyone, as even close friends have been known to react hysterically. While many people claim the right to be told, few indeed really need to know the diagnosis. Apart from the family doctor and health visitor (if appropriate), we advise parents to tell only those who are entrusted to look after the child such as regular babysitter, or nursery school teacher, in case events occur where incomplete medical information might jeopardize the child's care.

Initial responses to the news of a life-threatening illness include shock and disbelief. Where the mother is infected, guilt will be mixed with denial. Parents of children who were infected through receiving blood and blood products often blame themselves for agreeing to therapy which inflicted the illness. Denial and guilt may override the child's medical needs, in which case legal intervention is necessary. Some mothers focus on the child's symptoms to the exclusion of their own, while others use unrelated symptomatology to divert attention from HIV. We endeavour to ensure that infected parents receive counselling and appropriate medical care from adult colleagues.

As with most chronic conditions of childhood, parents are keen to join self-help groups. Sharing experiences and feelings, as well as anxieties, with other parents is effective in diminishing the sense of isolation and helplessness. With paediatric HIV disease, where so little is known about longer term prognosis, such self-help groups, are especially important. However, attendance at our mother and toddler group was poor, despite hosting it on neutral ground with crèche facilities in addition to providing both transport and lunch. The fear of breach of confidentiality was so great that the mothers could not bring themselves to attend the meeting.

Integration in the community

HIV is not highly contagious in the normal school or day care setting, as recognized transmission routes require sexual contact or parental inoculation. Studies of household contacts (including children) of patients with AIDS have failed to document horizontal transmission, and infection was found only in household contacts with known risk factors (Friendland and Klein 1987: Fischl *et al.* 1987). Children with HIV infection should therefore be allowed to attend nursery, day care, and school normally and participate in all activities to the extent permitted by their health.

In the United Kingdom, the Education Departments have accepted that there is no need for educational staff to know the diagnosis of HIV infection in a schoolchild. Children who are infected but asymptomatic will be unknown to medical staff as well as school staff. In areas of high prevalence of HIV infection, the staff should be stringently adopting policies and procedures to handle instances of bleeding. Even within an area of high prevalence, the risk of acquiring HIV infection from a single cutaneous exposure to blood from a schoolchild of unknown HIV status is minute. Hand washing and washing exposed skin with soap and water are the only mandatory precautions required, although gloves should be accessible for individuals who wish to use them. Under no circumstances should the child's care be delayed because gloves are not immediately available.

Children who manifest developmental delays, exhibit anti-social behaviour (biting, for example) or have extensive areas of weeping eczema merit especially careful consideration and discussions prior to a placement. In such instances, where special educational services will be required, the parents are advised that it is in the child's best interests for the diagnosis to be shared amongst professionals. Each child must be considered as an individual, taking into account his developmental status, clinical condition, and the expected interaction with his peers. Special education needs of such children may range from the provision of an extra member of staff to supervise the child who bites, to home tuition for the child whose health does not permit attendance at school.

Alternative care

The lifestyles of IVDU parents may be incompatible with child care; moreover, the parents themselves may be ill or dead from HIV infection. Some children have been abandoned in hospitals by parents. The cost of hospital care for children infected with HIV was recently published (Hegarty *et al*. 1988) where one third of the total inpatient days and over 20 per cent of the cost resulted from social rather than medical factors.

Consequently, many HIV infected children have had to be fostered or adopted. The Lothian Region Social Work Department has actively recruited foster families who are willing to care for symptomatic as well as asymptomatic children (Black and Skinner 1987). Some families have gone on to adopt the fostered child. It is hoped that in future these families will be able to provide terminal care for the child. Respite care may be necessary and could be provided by other carers or hospital staff.

Summary

AIDS presents a number of diverse and complex challenges to all child care staff. Health care personnel and statutory agencies provide an essential service, but voluntary organizations must not be overlooked as an important source of support to patients and families. In the UK, the vast majority of HIV infected children come from single parent homes where unemployment, imprisonment, drug use, inadequate housing, and low income already present problems prior to the diagnosis of AIDS. Young mothers and children can be concomitantly infected, and the cumulative cost of care far exceeds those of medical and nursing needs.

References

Black, A. and Skinner, K. (1987). Placement of children at risk of HIV infection. In: *The Implications of AIDS for children in care.* (Ed. D. Batty) British Agencies for Adoption and Fostering, pp. 36–7.

Brettle, R., Bisset, K. and Burns, S. (1987). Human immunodeficiency virus and drug misuse: the Edinburgh experience. *British Medical Journal*, 295, 421–4.

Fischl, M. A., Dickinson, G. M., Scott, G. B. *et al.* (1987). Evaluation of heterosexual partners, children and household contacts of adults with AIDS. *Journal of the American Medical Association*, 257, 640–4.

Friendland, G. H. and Klein, R. S. (1987). Transmission of the human immunodeficiency virus. *New England Journal of Medicine*, 317, 1125–35.

Hague, R. A., Yap, P. L., Mok, J. Y. Q. *et al.* (1989). Intravenous immunoglobulin in HIV infection—evidence for the efficacy of therapy. *Archives of Disease in Childhood*, 64, 1146–50.

Hegarty, J. D., Abrams, E. J., Hutchinson, V. E., Nicholas, S. W., Suarez, M. S. and Hegarty, M. C. (1988). The medical care costs of HIV infected children in Harlem. *Journal of the American Medical Assocaition*, 260, 1901–5.

Johnstone, F. D., MacCallum, L., Brettle, R., Inglis, J. M. and Peutherer, J. F. (1988). Does infection with HIV affect the outcome of pregnancy? *British Medical Journal*, 296, 467.

14 A social work service for cystic fibrosis families
TINA NEALE

Five years ago Barnardos was asked to provide a social work service for the families of children with cystic fibrosis who attended the Birmingham Children's Hospital. This has since developed to provide a community based social work service for all cystic fibrosis patients and their families within the West Midlands area, which covers Birmingham, the West Midlands conurbation and the the surrounding shire counties. Including about one hundred adults, there is a potential client group of over four hundred and fifty.

The work of the service

Five years ago a survey was conducted in which over one hundred families were asked what help they wanted and from whom. The areas covered in the survey included diagnosis and the information that had been offered, both at the time of diagnosis and subsequently; the financial implications of having a child with cystic fibrosis (diet, transport, heating, parental employment), emotional stress, spouse and sibling reaction, changes in social life, difficulties with baby-sitting or relief care, the routine of physiotherapy, and other immediate and long term problems, together with suggestions for practical solutions and help.

Five areas of need were identified which could be considered pertinent to life-threatening conditions of childhood in general and not specifically to cystic fibrosis.

1. Counselling is a relationship of trust within which advice is given, options are explored, and choices are made. This is required, not just at diagnosis, but through all major life events and including bereavement. The counsellor's innate ability, rather than his profession in life, is the most important factor.

2. Information about the disease should be available for families who would welcome more than is generally offered—always with the need for great sensitivity of delivery and timing. Information is always required in minority languages.

3. Better information is needed about the implication of cystic fibrosis for other professionals and the public at large. Poor treatment advice, well-meaning, but meaningless platitudes, hostile comments, and behaviour frequently result from ignorance. This is one area of need that has markedly improved over the last two years, following a number of television programmes and increased publicity from the Cystic Fibrosis Research Trust (see Appendix).

4. More practical help is required. Cystic fibrosis is a tedious illness. The routine is as much a problem as the skill and effort needed. Most families need help at some time and a few need help all of the time. By any standards some children are at risk because the parents are unwilling and unable to understand the necessity of the treatment regime.

5. Among a large minority, there is a need for more parental involvement in the decisions made by professionals concerning their children. This not only applies in matters of education and employment, but in those concerning choice of doctor, hospital, and treatment. The choice needs to be based on objective information. This may result in a move away from the consultant-led model of patient care to that based on a spider's web pattern with the patient and the family in the middle having access to a number of interconnected resources, all of which are of paramount importance at a particular period of time (Figs 10.1 and 20.4).

Barnardos Cystic Fibrosis Project

The Barnardos Project tries to meet these needs primarily through the work of the Project staff and a small number of volunteers.

Project Worker I has had considerable experience as a social work assistant and family aide. She visits and supports families who have difficulty concerning finance, isolation, deprivation, lack of knowledge, and lack of ability. She gathers resources, organizes holidays, arranges outings for single parents, and has become accepted as a surrogate 'mother' figure.

Project Worker II does not have a social work background. She has three functions. She recruits and trains volunteers to help directly and indirectly with families, she supports and initiates parent self-help groups, working closely with the Cystic Fibrosis Research Trust (see Appendix); thirdly, she visits families a few weeks after their diagnosis.

Project Worker III is a social worker who works directly with adolescents and young adults. She counsels them on any subject pertinent to move them on from paediatric to adult clinics. Working closely with professional colleagues, she is involved in work relating to transplants, pregnancy, dying, and bereavement.

The volunteer performs a number of tasks complementing the work of the professional. This includes befriending families, transport, baby-sitting, advising on welfare rights and helping to organize holidays.

All staff have a role in educating, collecting information and resources, and in bereavement counselling. The project has its own offices in central Birmingham in premises where meetings can be held, educational sessions organized, groups arranged with facilities for families to use on clinic day. These facilities are available for other groups who deal with the chronically sick child, for example, the Haemophiliac Mother's Group meets there. We have regular contact with the Birmingham Children's Hospital and some involvement in other CF clinics in the area. Referrals are accepted from any agency and self-referral is encouraged as we are trying to develop a pro-active rather than re-active service.

Barnardos has increased its commitment to this client group in three areas (see Appendix). A similar project has started in Nottingham for the East Midlands and is based at the City Hospital; three workers will provide a service to the cystic fibrosis patients in the area.

A variation of this service is developing at Leeds. The new project there also includes a social worker for other life-threatening diseases as well as cystic fibrosis.

In Newcastle, as part of the Barnardos Orchard Project for the North East, a single cystic fibrosis social worker has been appointed. He has joined an established project that provides a service for bereaved families.

The case for a specialized social work service

Social workers are the general practitioners of the Welfare State. They know a little about a lot of things and often a lot about a little, but nearly always there is a willingness to learn. Social work is about change—in attitudes, circumstance, environment, or behaviour. These changes can result from intervention in a number of ways.

1. Working with the individual. Methods used include giving advice, providing practical help, advocacy or using skills such as counselling to enable them to cope.

2. Working with the immediate social environment. A change in the attitude of the family, or school, for instance, will change the circumstances of the individual. Methods include the provision of information, family therapy, or helping with seconday stresses, such as finance, sibling behaviour or marital problems.

3. It is important to look at the existing resources available. If they are inadequate, or non-existent, then a social worker may be in a position to improve or provide. Help may be given with the organization and support of self-help groups, in developing a respite care service or in the recruiting and training of volunteers who can help families in numerous ways.

4. The system that affects the provision and accessibility of these resources in the first place may need improving or challenging. This is politics and Politics. Attitudes towards the chronically sick need to be changed by education, providing information to the public and the professionals, and in certain instances, by political lobbying and public awareness campaigns.

Chronically sick children as a client group are curiously undervalued by both statutory and voluntary organizations, and social work with sick children has a low status and low priority. Unfortunately, both national and local government policies dictate that other client groups have priority for available resources. The reasons for this are varied and complex and include a failure to recognize the needs of the very ill as distinct from the handicapped. Compared with other client groups, the dying have a low media profile, except when money is being raised which can perpetuate the myth that they are well catered for. Increased medical and technical skills have resulted in a rapid growth in survival rates, which in turn have outstripped the minimal services available. With the performance of the NHS being judged on quantity rather than quality of care, these children rely upon family and community services. However, community care is not a cheap option and there is a limit to the extent that any family can cope with its chronically sick members. It is also unlikely that there will be any profit in the privatization of caring for the dying.

Finally, there is a reluctance to be involved with such a painful experience as helping the dying, thus having to come to terms with personal mortality.

There has been an increase in activity among the child care charities and smaller medically orientated voluntary organizations to meet the needs of this growing group of children with chronic life-threatening disease, but this cannot compensate for cutbacks in the statutory services. Their expansion programmes too have been cut back by a fall in grants from local and central government. The increase in the numbers of charities means they are all dipping into the same pot. Anyway, should the care given to a dying child depend upon the proceeds of a jumble sale?

A chronically sick child is more than a medical or nursing problem. It is a spiritual, practical, emotional, social, and psychological one too.

Before dying there is a lot of living to do and it is in this area where the Barnardos social work service focuses its energy.

15 The Society for Mucopolysaccharide Disease and parent self-help groups

CHRISTINE LAVERY

The number of self-help groups for families of children suffering from life-threatening conditions increases continually to keep pace with the demands of parents seeking an understanding of the medical and social needs of their children, as well as combating the immense feeling of isolation that accompanies the diagnosis of a rare disease. All doctors are aware of cystic fibrosis. If you mention this disease to your neighbours there is a good chance they will have heard of it. For each of the better known life-threatening conditions affecting children there are many hundreds of which most doctors—let alone your neighbours—will never have heard.

The Society for Mucopolysaccharide Diseases started with a very small nucleus of affected families and has seen its membership grow to three hundred and fifty families with living affected children, and a hundred bereaved families. In 1976 when our son Simon was diagnosed as suffering from Hunter disease, my husband and I were told we were the only family in the country with a Hunter son. We learnt firsthand the feelings of isolation and ignorance. Simon died at the age of 7 years, just two months after we met another Hunter family.

Thankfully, seven years on, no newly diagnosed family need love and care for their child or children who suffer from mucopolysaccharide (MPS) disease in such total isolation and ignorance. The Society has established a network of Area Families who offer support through personal visits, and organize family activities at local level. The Society holds a National Family Weekend Conference each September, attended by at least one third of the families. They come from every corner of the British Isles and beyond, hungry to expand their understanding of the diseases as well as share experiences with similarly affected families. From the outset we recognized that these weekends must cater for the needs of the MPS family; mother, father, affected children, and brothers and sisters.

Over the years we have trained a team of sixty volunteers, between the ages of 16 and 65 years, to help with the children. They come from a variety of backgrounds, but what they have in common is profound understanding of the needs of MPS families. Whilst parents participate in the conference and workshops, the children are treated to the delights of a theme park. The under 10s are tucked up in bed watched over by a volunteer, the parents settle down to enjoy a dinner and disco; the talking goes on. A theatre group encourages the teenage MPS children to take the stage, whilst their brothers and sisters let off steam with a pool party. The benefits of this annual event can be summed up in the words of one mother; 'My batteries have been recharged for another year'. The weekend is one of many activities heavily subsidized by the Society, and in cases of financial hardship grants are given to enable a family to attend.

As well as bringing about awareness of these rare diseases, the Society publishes specialized booklets on individual MPS conditions, a quarterly newsletter, and an annual conference report, all of which are in demand by families and professionals alike.

We are always looking for new ideas to support MPS families and for the past three years have organized a group holiday for fifteen affected families. Our holiday co-ordinator, plus a small handful of volunteers, accompany the families, providing added support as well as having the energy to organize outings and babysitting. Families of healthy children can all too easily take for granted annual holidays and weekend breaks. It is a different story for families of children with a life-threatening condition. The energies that go into planning special events cannot protect against the sense of disappointment when the holiday has to be cancelled because their child is too ill to take part. The unpredictability of children with life-threatening conditions leads many parents to live from day to day.

Because of the life-threatening nature of MPS and the fact that there is no cure, families are keen to keep abreast of any research developments and raise large sums of money for this purpose. Perhaps the families' greatest achievement to date is the appointment of a Consultant Paediatrician at a children's hospital specializing in the mucopolysaccharidoses and mucolipidoses. The Society was involved in the development of this post and the subsequent appointment. It negotiated terms whereby the Society provides all the funding in the first year, three-quarters in the second year, and half for the third and fourth year. Thereafter it becomes an NHS post, leaving the Society free to look at other areas of much needed research.

A growing number of doctors and professionals in the field are recognizing the value of parent-orientated, self-help groups. Presenting a life-threatening diagnosis to a family must be one of the most difficult tasks undertaken by a paediatrician. How this is carried out will have long term implications for the family. I do not suppose the doctor who put a handwritten note through a family's door saying, 'Thought you'd be interested to know your child has Sanfilippo syndrome' will ever read this book, but perhaps it will be a lesson to others that bereavement starts with diagnosis. I am sure that the same doctor would not put a note through one of his patient's doors saying 'Thought you would be interested to know your husband has just died'.

Many doctors still shy away from giving parents information on any relevant self-help groups. In addition to my work for the MPS Society, I serve as National Development Officer for Rare Handicap Groups at Contact a Family (see Appendix). I receive over a thousand enquiries a year from distraught parents trying to find another family in similar circumstances. In some situations the child's condition is so rare that we have to look beyond the United Kingdom for help. However, for most enquirers a self-help group exists. Some are just a handful of families in touch, others are more established. What all have in common is the ability to let families know they are not alone, offering a shoulder to lean on, sharing experiences on the management of the child, and on the organization of family life.

At Contact a Family we are closely in touch with all the established self-help groups and have set up a training programme in twelve locations nationally, for parents running self-help groups. We hope that it will not be too long before a group exists for every condition, however rare, able to offer support to families involved; equally, that many more professionals will give recognition to the role of the 'professional parent'.

Part V

Education

16   *Enhancing their lives: a challenge for education*

PETER JEFFREY

Being a child and going to school are seen as part and parcel of the same thing by most people. Schooling is generally supposed to prepare children for adult life and the future; important though this preparation may be, school is in itself a major part of a child's life, here and now. Among many other things, it is a place to meet other children and form friendships; a place to develop skills and experience discovery, achievement, and creative expression. Thus to answer the sceptics, there is every point in schooling for children who may not survive into later life.

It has to be recognized that schools are places filled with large numbers of lively, active children where it is not always easy to consider individuals with particular needs. This represents a challenge to teachers to take account of an individual child's special circumstances, without making their pupil feel too different. In this way skilful teachers may succeed in educating children with life-threatening conditions within the normal environment of ordinary schools.

The question is not whether a child has a serious illness but how much and in what way it affects his or her ability to benefit from the schooling provided. The Education Act 1981 lays down procedures for assessing special needs if it seems likely that a child's education will be adversely affected. Once these needs have been assessed, a local education authority must make appropriate arrangements to meet them and keep them under review. A range of options may be available, from an ordinary school with the necessary additional aids, adaptations, and assistance, to a special school with its concentration of resources, to teaching at home or in hospital. The result of an assessment may be that the education authority issues a formal Statement of Special Educational Needs which should guarantee the necessary provision. However, in many instances of life-threatening disease, this will not be necessary.

Unlike most chronic and non-progressive handicaps, many life-threatening illnesses vary considerably in their effects from time to time. For this reason, an education service should aim to respond flexibly. In Waltham Forest, a service has been established for several years for children with long or short term sickness or physical disability, which brings together support for ordinary school attendance, long or short term special school placement, and home and hospital teaching. These services are operated from a single base at the special school, allowing a variety of responses to individual circumstances.

An adaptable approach can only work well when it is organized in such a way as to provide educational continuity. This means that all the elements of a service and all other agencies involved need to be closely linked so that any recommended changes in the arrangements for a child's education can be made without delay and without undue formality. The essential ingredient in an education service's response to a child's changing needs is close co-operation and agreement with parents and, if appropriate, the child.

As teachers are so rarely called on to work with children affected in this way, they have usually worked in isolation. With the kind of provision described here, the experience of the few teachers working with seriously ill children is focused on one place, and has the effect of creating better awareness and assisting the development of good practice. The unified service links together teachers in hospitals, at home, in ordinary and special schools, and is able to pass on the benefit of accumulated experience to those who may be faced with this situation for the first time. The service works to raise the awareness of all teachers in the area by arranging occasional meetings, inviting specialist speakers to explain something of the conditions involved and the effect of symptoms and their treatment on the child in a school setting. These lectures are attended by community health staff as well as education staff, giving teachers, clinical medical officers, school nurses, and health visitors a chance to learn something of one another's concerns, responsibilities and resources while maintaining a focus on the effects of different conditions on the lives of children and their families.

This indispensible co-operation locally between the education and health services has given rise to the appointment by the Health Authority of a specialist nurse who is based with the service working closely with its specialist head teacher and other education staff. The specialized nurse has an important part to play in helping to plan for the continuity of an individual sick child's education, acting in general as a point of contact for the doctors and the community health service and education service to sick children. This makes it possible to offer parents suggestions and choices for their child's schooling in the light of a shared understanding of the child's needs. It is the responsibility of the unified service to liaise with the headteacher of the child's school, the school psychological service, the community health service, and the education department. Without this facility and spirit of co-ordination, arrangements between so many diverse individuals and agencies are likely to move at too slow a pace for an individual sick child's changing needs. The overriding consideration is to open the way for the best kind of educational opportunity for the child.

There are those who think that if every cloud has a silver lining, for the seriously or terminally ill child it is that at least they do not have to be bothered by schooling. Our belief and experience would argue otherwise. The most reassuring thing to children with life-threatening disease is to know that they are expected to live their lives as others do. Removal of the normal challenges, disciplines, and expectations of school life can be worrying; similarly unlimited treats and uncritical acceptance of misbehaviour can cause anxiety.

A teacher working with a terminally ill child at home requires great skill, sensitivity, and judgement to play several parts. Although a schoolteacher's job in general is seen as being the education of children, the actual work of individual teachers varies widely: some work with under-fives, some teach a broad range of subjects at junior school, others specialize and teach a particular subject to an advanced level, while others take responsibility for careers education or pastoral care. Teachers are expected to provide discipline, motivation, care, and counsel. They may be called upon to work with children who have special needs, who have behavioural, learning, or physical difficulties. A teacher working at home may be required to offer an even wider range of skills than would be expected in school. They may not always find they are completely prepared and need help in making judgements about what to offer a seriously sick child at home. The teacher really needs to know the answers to certain questions. What is the outlook for the child? Are all members of the family—parents, brothers and sisters, and the child—aware of this? Without knowing the answers, a teacher will have difficulty in knowing what to offer or indeed how to relate to their pupil.

As with the kindly-intentioned indulgence of sick children already mentioned, the totally undemanding teacher may convey the message to the child that there is no need to make an effort to do any more. It would be wrong, and indeed impossible, to generalize just how the teacher should behave; this is why sensitivity and judgement on the part of the teacher, as with all adults concerned with the child, are so important.

There can sometimes be conflict over the demands of formal academic work such as that required to follow a public examination syllabus. The fact that a child may choose to work towards an exam does not make the work any easier; the academic stresses and strains of meeting the coursework requirements will still exist and be shared with healthy children. Coursework, project work, and homework will cause the usual student worries. But the very normality of these demands can help children feel a part of life and divert some of their anxieties about symptoms and outlook into an area where they have a chance of overcoming obstacles by their own efforts. All of this assumes that academic pressure is tolerable within the limits imposed by an illness and its treatment; this calls for detailed discussion with medical advisers. It is observable that in spite of the usual adolescent protest, children themselves rarely want to withdraw from examination groups or courses. More often it is parents who feel protective and sometimes angry at what is seen to be an unnecessary imposition on their son or daughter. This emphasizes the need for partnership between parents and teachers, taking proper notice of a child's own feelings and capabilities. Examination work shows this up most clearly, but the same considerations apply to school work generally.

Sometimes a seriously ill child's ability to do things will be affected as the condition progresses. Deteriorating hand skills, vision, speech, or concentration are examples of this. To some extent these problems can be overcome, keeping children interested and active in learning and creative activities through the use of technical innovations. A great range of equipment, especially computers, now exists which enable children with impaired functioning to go on producing results from their efforts, something which was barely possible a few years ago.

For children whose illness is causing pain, reducing their strengths and morale to the point where it is unreasonable for them to expect to keep up with everyday demands, it is another matter. The teacher's role at times like this is, if possible, to encourage a more passive engagement in listening to, or looking at, what is presented. There are, of course, times when withdrawal of lessons and attempts to occupy the child is more appropriate, and indeed, times when an illness allows no choice.

Many people's experiences suggest that children who are dying often know or suspect the truth whether or not this has been shared with them. Awkwardness, confusion, and, above all, inconsistent answers to questions by adults coming into contact with dying children possibly generates greater feelings of insecurity than a supportive sharing of the reality. Adults have a responsibility to try to work out how to tackle this dilemma in the best way for individual children. Close family members will need to be consulted, and indeed supported, in reaching the right decision for their child; this may provide a consensus and climate in which to provide the communication the child may wish to make. Even then, a child may choose not to ask about or invite talk about death; some children would appear to prefer to protect others from the distress of such a discussion, particularly their parents. It is certainly not the job of a teacher, nor perhaps of anyone but the child's parent(s) or closest confidant to open the subject for them.

The benefits of medical science for these children offer the opportunity for a full and worthwhile if limited life, not simply the alleviation of symptoms and the postponement of dying. Of all those associated with sick and terminally ill children it is perhaps the teacher who has the greatest opportunity and responsibility to enhance the quality of that life.

17 The Macmillan Education Centre: a resource for professionals caring for dying children

ANN DENT

Introduction

In 1967 Dame Cicely Saunders opened St Christopher's Hospice in London. This heralded a new approach to terminal care, which has now spread throughout the country and much of the world.

Many people dying of advanced cancer since then, have benefited from those original concepts which still hold good today. We now appreciate that a dying person is a living person; not just a disease with a failing body. The afflicted individual suffers emotional, spiritual, and social distress as well as physical pain. We have come to recognize that the needs of the family are also important, both during the illness and in bereavement, and that a quality and normality of life should be encouraged wherever possible for the whole family.

Nurses more than any other profession have begun to understand and practise these concepts. It has meant changing our role from being directive to being non-directive, or learning and practising good symptom control, and using our imagination to meet the needs of individuals and their families by patient listening.

Change is nearly always a painful process and the burgeoning of the hospice movement has proved no exception. In the early 1970s, the advantages of hospice care began to be recognized and appreciated. Many groups throughout the country were keen to emulate St Christopher's. However, to build, equip, staff, and maintain such a building was extremely expensive. The Government cut-backs of 1973 made it impossible for Health Authorities to contribute to local hospice charities. In 1977, the first national purely domiciliary service for terminal care of adults was started at the Dorothy House Foundation in Bath. It soon became apparent that this was not just a cheap alternative to a hospice building but a complementary and effective way of caring for dying people and their families.

Understandably, the development of a service of home care hospice nurses produced feelings of disquiet among community nurses. District nurses saw their role with dying patients being taken away from them. However, as a result of widely available education and information the concepts of the hospice movement are no longer the sole right of those working in hospices or home care. It is hoped now, that with awareness and knowledge, that all caring professionals will practise good terminal care within their own settings. Specialist hospice nurses should be used as a resource, a fact which Cancer Relief (see Appendix) has recognized by supporting Health Authorities to employ Macmillan Nurses (see Appendix) as a specialist back-up to existing services.

It has been largely through education that the quality of care for dying adults has changed. Fortunately, the death of a child is rare, but it is because of this rarity that few people have experience or understanding. As a result, it would appear that there is little education in the country focusing on the specific needs of the family where their child is dying; likewise there is scope to pay closer attention to the needs of those nurses and other health care professionals as they support families in providing paediatric terminal care. Because of this apparent lack it was agreed that the Macmillan Education Centre in association with the Dorothy House Foundation, Bath (see Appendix) sponsored by Cancer Relief, would support an educational resource.

Macmillan Education Centre

The Centre opened in January 1988. It has four main aims:

1. To offer continuing education to Macmillan Nurses working in the community and also to their managers.
2. Likewise, to offer continuing education to Macmillan Nurses who are hospital-based.

3. To offer an educational resource and support to other health care professionals who are involved with such care.
4. To evaluate the provision of this educational resource and to make recommendations for future development.

Paediatric terminal care

The ultimate aim is to give families the opportunity of caring for their dying child at home, supported by competent, confident, sensitive members of the primary health care team, working closely with the hospice team, from the time of diagnosis through the illness, into an advanced stage disease if there is no cure, and beyond death in bereavement. We do not aim to run specialist courses in oncology but to help professionals to use the concepts of the hospice movement when cure is not possible.

When a child is dying from any life-threatening disease we believe that community nurses, especially health visitors, are in a position to support the child's family from the time of diagnosis. So, our aim must be to address the nurses' need for education in this area.

Two-day workshops are now being run on a regular basis, entitled, 'Caring with confidence', the aims of which are:

1. To encourage confidence of the trained nurse in working with families where: (a) a child has advanced stage disease; (b) children may be involved when an adult is dying.
2. To help hospice and community nurses to work more closely together.
3. To help nurses understand more fully the grieving process in parents, to increase self-awareness of their own attitudes towards death and dying, and to enhance counselling skills.

Workshops have not only been held at Dorothy House, Bath, but throughout the country, where professionals have recognized the need to develop their skills. Many of these have been arranged in conjunction with the Lisa Sainsbury Foundation (see Appendix), an organization set up for professionals to promote better understanding of terminal care for all ages. All Macmillan Nurses who attend courses, either at the Dorothy House Foundation in Bath, or at Basingstoke (for hospital-based Macmillan Nurses) participate in a session on 'Children and dying'. We have also run day courses for the clergy on 'Death and the family' concentrating on children who have experienced bereavement.

The first residential, multi-disciplinary course was held in December 1988, near Bath, when forty professionals attended from all over the United Kingdom. As a result of that conference, a small group of clinical nurse specialists now meet bi-annually to discuss their roles and to consider future developments of the service they offer. A workshop on 'Teaching skills' for these nurses was held in November 1989 with a view to encouraging them to run sessions for professionals in their own area. In 1990 seminars are planned for teachers in primary and secondary schools, a three-day course on bereavement of children or parents, together with day workshops which include one for play leaders in hospitals, another for hospital-based teachers. It is hoped that in the future there will be more multi-disciplinary workshops for professionals, to learn from each other and to respect each others roles, as a move towards better team work. At the end of October 1989 a professional pack became available entitled '*A child is dying—care of the family during illness and bereavement*' (Dent, 1989). Available from the Macmillan Education Centre or Meditec (see Appendix), it is primarily produced for nurses, using the hospice concepts and giving simple guidelines so that they may care for families with more confidence.

Until recently it has been difficult to find books relevant to the subject. However, Meditec has now kindly produced a special book list for us entitled, '*Children, the family and death*'. A synopsis of all their books is available in the new pack.

Conclusion

The field of paediatric palliative care has much to learn from the adult hospice movement. We should learn from its mistakes and concentrate on the positive aspects. As we have seen, home care has been shown to offer families a valuable dimension. We believe that for a family caring for a child who is dying, this is even more important. However, there must be on-going local supportive care throughout the illness and beyond. To achieve this, it is important that members of the primary health care team are given the opportunity to explore their fears and anxieties and to understand the needs of the family so that they may feel more confident to care. We aim to provide this facility, albeit in a small way, but hopefully the ripples will spread through education and understanding, and more families will be cared for in what surely must be one of life's most traumatic experiences.

M.O.T.

You can keep a car for years
Or just a few months
Depending on how you treat it,
Me, I've failed my M.O.T. twice
But I'm still running.

Tracy Wollington

Part VI

Bereavement

The Alder Centre: counselling for all those affected by the death of a child
OWEN HAGEN

The Alder Centre is the outcome of bereavement support work started in the early 1980s by a number of professional groups within the Royal Liverpool Children's Hospital and our social work department in particular. This work is continuing and developing and it will be nurses, doctors, and social workers who will sustain the major part of the task to care for those left to grieve after the death of a child. However, these professional carers will now have the additional support of the Alder Centre, a project which they have developed and which will in turn help sustain them.

The Centre is within the grounds of the Royal Liverpool Children's Hospital, Alder Hey and is housed in a specially designed suite of rooms away from the main hospital building. With its two full-time counsellors and a secretary, and with accommodation which allows for both individual and group counselling, it both acknowledges and extends our commitment to helping the families of those children who died at, or are brought dead to the hospital. It will be there for *all* others affected by the death of a child—relatives and friends as well as professionals in the hospital and the community at large.

Experience has taught us that good terminal care does not simply 'happen'; it is the product of careful thought by, and good communication between, experienced staff. Likewise, support for bereaved families needs the considered response made possible by a co-ordinated service. Families should not be left until they are desperate and have to ask before they are offered help. Bereavement support has to be organized and actively offered in a manner which allows families to make a choice. How many bereaved parents (some still in their teens) will take up the offer to, 'come back any time and talk if you have a problem' or will respond to the well meant gesture, 'you know where we are if there is anything we can do'? Parents who are shocked, disbelieving, angry, or racked with doubts or feelings of guilt will make few demands on us as their world seems to close in on them. Undermined by the enormity of the event and with self-esteem badly shaken, they will offer appreciation for the care given and silently acknowledge that the hospital's main concern is for the living; isn't that what it's there for?

It is worthwhile considering the circumstances which allow parents to feel abandoned by the hospital and by staff who have cared so attentively for them and their child. The family may literally have lived on a unit for intervals, over several years, nurturing friendships and the kind of closeness that comes from working through situations of crisis. Some feelings of loss and rejection are inevitable as medical care is withdrawn. But how can staff who rightly take pride in looking after the whole of the family, showing enormous kindness and concern for a grandparent or parent or brother and sister, fail to extend this to the period after the child dies? Do the family's needs disappear? Is it that staff don't care?

Hardly. The answer has little to do with care and everything to do with the division between community-based services and the hospital. There is a need to bridge this gap and identify who is best placed to offer support, and then allow them the means to do so; of equal importance are the feelings of helplessness workers have to cope with when faced with someone who has lost a child. Generally we are more comfortable when we are doing something that has a tangible and positive outcome—being seen to provide a service, meeting a specific need. The Alder Centre is there to help all workers acknowledge and tolerate their inability to give the bereaved what they really wish for.

What then can we offer bereaved families, and how can it possibly compensate for the enormity of their loss? The answer is that nothing can—and nothing should. At first sight this seems to leave the individual worker with very little other than the collective and individual distress of those affected by the loss of a child; but there *are* things we can do, some simple, some less so, some of which might significantly alter the capacity of the bereaved to cope with the experience. It is worth listing some of the resources which parents have found helpful:

1. *Practical help and advice in the days that follow the child's death.* This can range from welfare rights advice to information about the necessary procedures which are taking place (post-mortem, inquest, registration of death, funeral arrangements), or organizing the attendance of a representative from a particular religious denomination.

2. *Information about the cause of death, circumstances of the child's death, treatment offered by emergency services, and medical care given* (which is immediately of great concern for some parents, although for others this is needed at a later stage). This most helpfully takes the form of an interview, or interviews, with a consultant paediatrician, in the family home if necessary. Some parents may value assistance in preparing questions beforehand. It is our experience that this interview can be a landmark in helping parents release the burden of responsibility for their child's death, which is commonly felt, however irrational that might seem.

3. *Individual counselling* and help offered by any one of a range of professionals—this will be determined by individual circumstance, and to some extent, by the confidence which the worker has in their own capacity to help. Counselling may come from social workers, health visitors, voluntary agencies, or the clergy as long as problems of overlap are avoided.

4. *Access to another parent who may have suffered a similar loss.* A well established network exists in this hospital for those who wish to be put into contact with another bereaved parent.

5. *A number of bereaved parents' groups meeting at the hospital,* both for families which have lost a child suddenly and for those which have lost a child following an illness.

Circumstances and parents' pre-occupations obviously differ between the groups, although the strength of feeling present is equally tense.

6. *Parents are welcomed back to the hospital if they have anxieties about either subsequent or surviving children.* A clinic has been established for parents who have had a cot death and intend having another child, addressing the particular anxieties these parents experience.

7. *The Alder Centre is a point of contact for families who may need specific help at a difficult moment, possibly some considerable time after their child has died.* Some families will feel that they

have reached the stage at which they can help others by drawing from their own experience and their efforts to place meaning on such a terrible event.

These are very real resources and have been created by staff committed to supporting the bereaved. They form the backbone of a service which did not exist ten years ago.

When in 1984 a group of staff called an open meeting in the hospital to form an *ad hoc* group to review the hospital's terminal care and bereavement support provision, we took a step which had far reaching consequences and which pointed to the need for the Alder Centre. We invited bereaved parents to join the group. The Alder Centre is a product of this unique partnership. Parents are present on the organizing committee, at interviews, and shortlistings of counselling staff. They run their own support groups alongside those organized by social workers and nurses and produce a regular newsletter. They raise large amounts of money from essential fund-raising activities, (the other major sources of funding are Liverpool District Health Authority, the Nuffield Provincial Hospitals Trust, and local business and commerce). They also liaise with the media, producing display and publicity material; run parent groups in their own homes and above all, support one another through the appalling experience they all share. This partnership has given the project a dimension with few equivalents within health care, as well as an irresistible momentum.

The message then from parents is, 'don't hold back'. The bereaved are normal people who have experienced a terrible event. They are not to be pitied or shunned. They come from every social class, ethnic group, and religious faith. Every personality trait is represented, every quirk and failing; the broad-minded and the bigot, the depressed and the cheery, the silent and the ebullient, some embittered and others constructing new meanings and purpose. They are our colleagues and our neighbours, they may be you, the reader. In some respects their needs are simple and are felt in differing ways, at certain times in life by all of us: the space to dwell on the person who is absent and express the depth of pain being felt, to be listened to and to be allowed to speak, to obtain what information there is about why and how the circumstances have occurred. What separates out the experience of losing a child from other traumas is, per-haps, the depth of despair in those left to grieve, the fear of being overcome and destroyed by the enormity of it all.

The Alder Centre thus hopes to address the many concerns of families following bereavement including: support for staff working with very ill children, something often talked about but notoriously difficult to facilitate; help for children affected by the death of a brother or sister, parent, or person of import-ance to them; training for hospital and community based health care staff. A book of remembrance will be kept as well as an album of photographs and family snaps. Referrals will be taken from the general public, regardless of the time which has elapsed since the death.

The Centre will be a focal point, a place to turn to, a place in which to grieve and possibly reconstruct life following such an overwhelming loss to place a meaning on the death, perhaps through helping others who have suffered a similar loss.

19 *The Compassionate Friends: mutual support by bereaved parents*
JILLIAN TALLON

This book is mostly about caring for children who are ill and the support offered to their families. For some, perhaps for many, death and bereavement are the outcome of such illnesses.

When a child dies, it is the parents—and, sometimes, especially the mother—who are the main focus of care, but it is tremendously important to include the wider family, and most particularly the sisters and brothers. The parents will need help individually and together—women and men tend to grieve somewhat differently from each other—the other children will need help, and often the grandparents too. There are few grandparents who would not rather give their own life than see their grandchild die, and their grieving and suffering are acute. They suffer the loss of their grandchild, they witness their bereaved son or daughter's grief, and they may endure 'survival guilts' themselves.

It is difficult to write in 'general' terms about grief of brothers and sisters because the term embraces such an enormously wide age range. Very young children cannot understand what has happened, but must be acutely aware of the unhappiness and stress that is around them. As they get older their understanding increases but still they need to have things repeated many times. Because you have explained, at whatever level they can comprehend, that their brother or sister has died and they seem to have understood this does not necessarily prevent them asking for their sibling again next day—and it can be excruciatingly painful for the parents or older brothers and sisters, or grandparents to have to re-explain it all. Truly a rubbing of salt in the wound, but something that cannot be avoided.

Books can be helpful in conveying what we are talking about, and books are available for all ages of children. They portray death through different 'vehicles'—animals, pets, the cycle of life in plants and trees. There are books about the loss of grandparents, and increasingly about other family members and about friends.

Bereavement projects are opening in many towns across the country, sometimes springing from religious organizations, sometimes from secular backgrounds. The number of support organizations to help people cope with bereavement is increasing. There is Cruse, which offers counselling support through all types of bereavement (spouse, partner, child, parent). There is the National Association of Bereavement Services, which aims to become an umbrella organization for very many of the bereavement services available across the United Kingdom. These groups are generally counselling organizations. This implies a rather more formal, structured kind of relationship, usually by appointment for an hour on a weekly, fortnightly, or monthly basis. Additionally, there are organizations such as Compassionate Friends, the Foundation for the Study of Infant Deaths, the Stillbirth and Neonatal Death Society (SANDS), the Twins and Multiple Births Association (TAMBA), or the Miscarriage Association. These are befriending organizations where bereaved parents and families are put in touch with others who have endured a similar bereavement and where, generally a much more informal relationship—friendship—is offered. Contact may be on a one-to-one basis, by letter, by telephone, through daytime or evening meetings, and occasionally at weekend meetings (see Appendix).

The two kinds of support, counselling and befriending, are not 'mutually exclusive'. There is no question of choosing one *or* the other. They are complementary not competitive, and there is plenty of room for co-operation in the most literal sense of the word, of working together. Many parents will benefit from *both* types of help, drawing strength and encouragement from the befriending organization; while through the counselling they come to an understanding of their personal responses to the child's death. The hideous shock each morning as you wake up and have to re-remember that it *is* true, that Robert or Jennifer *is* dead, that it wasn't a terrible nightmare. The guilt that strikes a parent the first time they laugh after the death of their child! The irrational but very real sense of betrayal when you look out of the window and think 'What a lovely morning'!

But very slowly—and whether we want it to or not—the death gradually becomes an accepted fact, and the painful task of moving back into the 'mainstream' of living is begun. It is slow, it is uneven, and, on a day-to-day reckoning, we are aware of little change or development. But when we look back over, say, six months we can see that the integration is taking place. We will never forget our child. It is impossible. They were, and are, and always will be, a part of us. But gradually the raw agony of remembrance eases, and slowly we begin to enjoy remembering our son or daughter, that they were part of us and part of our family, and they continue to be so. They are included in conversations, photographs are looked at and the events surrounding them discussed. We don't stop missing our child, ever; we don't stop wondering how they would have developed, physically, in character, and as personalities, of career choice and their own family perhaps. We don't truly 'accept' their death, but we do learn to live with it.

The Compassionate Friends

The work of most of the befriending organizations mentioned above is explained in their titles. The work of The Compassionate Friends (TCF), however, is less obvious. This is an organization of bereaved parents—everyone working within TCF has experienced the death of their child (and sometimes children). The words 'parent' and 'child' know no age limits, and 'death' includes all causes—acute, long term or genetic illness, accident, murder, and suicide. With its office in Bristol and a network of county secretaries, TCF is a nationwide (and registered) charity, founded in 1969. A range of leaflets is available to parents and families, to friends, and to the breadth of professionals involved in the death of a child (including one called *Preparing your child's funeral* and another entitled, *Bereaved parents and the professional*)

Support is offered to parents in a number of ways, both locally and nationally. The quarterly newsletter is sent out from Bristol —it is a lifeline for many parents, especially the newly-bereaved. They draw strength from finding out that they are not alone, that others have experienced the same *symptoms* of grief, that there is a glimmer of light at the end of a seemingly endless tunnel. There is a postal library of some 400 titles, for use by both parents and professionals, and a booklet called *Helping ourselves* which many parents find very helpful. TCF adds to its range of leaflets at intervals. There is a 24-hour answering service, with a referral number for emergencies.

Perhaps the support that is most valued by parents is the opportunity for one-to-one contacts and group meetings. Meetings may be morning, afternoon, evening or, occasionally, weekends. The meetings vary greatly in their structure and content, according to the county secretary or group leader. Some function as 'tea and sympathy'—and this is in no way to decry that provision; it is immensely therapeutic to have somewhere *safe* to cry, where people don't try to make grieving parents 'feel better', where others will cry with them, but will also help them mop up at the right time, and encourage them to keep going. Others groups may be more structured, sometimes with an invited speaker. Some make use of video or audio cassettes. Parents are generally put in touch with their county secretary, or nearest geographical contact, but many also appreciate the opportunity to share with someone whose child has died from the same cause or in similar circumstances. Every effort is made to link such parents, when such contact is asked for. There are contacts for grandparents of 'both kinds'—those bereaved of their grandchild, and those whose adult child has died, leaving grandchildren. Attempts are being made to offer support to brothers and sisters too. Within TCF are two special groups whose members share the same grief and difficulties as everyone, but bear extra burdens in terms of the law and the media. These are, The Parents of Murdered Children, and The Shadow of Suicide. TCF is an international organization too, with contacts in many countries.

Each of the organizations mentioned (and many additional ones) will do all they can to help bereaved parents. None can provide the one thing that every parent wants—their child, back again, whole and happy. The befriending organizations offer the support that comes from having experienced a similar tragedy, and somehow having come through it. Some can identify what has helped them do so—it may be their faith, their family, their job, a particular friend. A lot of people will say that 'The Compassionate Friends (or The Foundation or SANDS or . . .) helped me to believe that I could and would make it, even when I didn't think I could, even when I didn't really want to'.

Wave goodbye

I don't regret a single moment,
Of my hate and dreams,
But will I regret the hurt
As I wave goodbye.

Life seems so long
Yet I have no time,
Life has escaped my hold
Moved too fast,
It was always just
A matter of time,
Before I waved goodbye.

It's hard to smile with false intentions,
My face laughing blankly,
Instead inside I feel
The tears are building
Soon I will drown them all
In the flood
Lift my hand, wave goodbye.

                                    Tracy Wollington

Part VII

Overview of needs and services

20 Towards a comprehensive system of care for dying children and their families: key issues

ROSEMARY THORNES

The working party on the 'Care of dying children and their families' was set up by the National Association of Health Authorities (NAHA), the King's Fund, and the British Paediatric Association in 1987. This chapter introduces the report produced by the working party (Thornes 1988) and concentrates on some of the key issues. Its brief was clear: to provide guidelines for health authorities on the care of those groups of dying children for whom plans could be made. The gap in policy, provisions, and knowledge of good practice for this client group had become increasingly obvious after the publication of *The care of the dying* by NAHA (Eardley 1986) and the DHSS Health Circular HC(87)4 (DHSS 1987), both of which dealt mainly with adult care.

During 1986 and 1987 several conferences and seminars were held, concentrating on the special needs of children (Salvage 1986). These began to define possible elements of a comprehensive service, but many uncertainties remained. The success of Helen House seemed to deepen the confusion. Mother Frances urged restraint; perhaps what was needed were not copies of Helen House but rather a translation of its philosophy to more local, community-based services (Dominica 1987).

The experts were failing to convey the message that children were different and needed a distinct type of care. The general public, misled perhaps by the word 'hospice', seemed to presume that most children were dying of cancer and needed facilities similar to those provided for adults. Moreover, no-one knew how many families were involved!

It was against this background that the working party started its deliberations. To streamline their approach to this confusion

of issues, a decision was made to start with the children themselves and then follow a simple flow diagram (Fig. 20.1) in
the collection of information and the ensuing discussion.

Fig. 20.1   Organization of the discussion and the final report

These headings now form the main sections of the report
(Thornes 1988), with additions showing where to turn for
further information and advice and a summary of the main
recommendations. The aim was to produce a practical report,
providing guidance for individual practitioners and at the same
time enabling health authorities to see their next step, encouraging them to work towards a comprehensive service.

Who and how many

These guidelines do not deal with children who die unexpectedly; they are concerned with those for whom plans can be
made and appropriate care offered. These children can be separated roughly into three groups:

1. Those with progressive degenerative conditions for which
   there is no current cure, such as muscular dystrophy, mucopolysaccharidosis, or cystic fibrosis.

2. Those with a handicap which is so severe as to be a threat
   to life.

3. Those with a life-threatening disease, such as leukaemia, liver
   disease, or a profound heart malformation in which medical

intervention may prove successful, but which still cause a sizeable number of deaths in childhood.

We have estimated that at any one time there is a minimum of 5400 children in the UK who will either die in the current year or who already have the threat of death hanging over them. In an average health authority, at any one time, there are likely to be:

- 2 who will die from cancer within the year;

- 1 or 2 with major organ failure who will die within the year;

- 4 who will die within the year of a degenerative disease or life-threatening handicap;

- 16 at an earlier stage of a major illness, a progressive degenerative condition, cancer, or life-limiting handicap, whom we assume will live some three years, but who are not yet in their last year;

- 3 with cancer within a year of diagnosis, who will survive;

Thus 26 or 27 families are involved.

The family

The working party's first action was to add 'and their families' to their original title. There was unanimous agreement that a child should never be considered in isolation, but always as part of a family. Throughout the report, the role and needs of all the family, but especially the parents, are stressed.

The family is centrally involved. Home is the centre of caring and one of the objectives is to keep the family together. The parents have to make decisions and need to maintain control of the situation. To these ends, any input to care from any professional or agency, should be based on an honest approach and the sharing of information. Families also need emotional and spiritual support, adequate and appropriate support in the home, and often extra finances. In return, parents will produce an unmatchable expertise in the care of their child and many will build up a more detailed knowledge of the disease and treatments than is available elsewhere in the district. (Fig. 20.2).

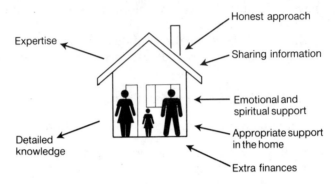

Honest approach

Expertise

Sharing information

Emotional and spiritual support

Appropriate support in the home

Detailed knowledge

Extra finances

Fig. 20.2   Maintaining the central involvement of the family

Continuity in support and care

For parents bereavement starts at the time of diagnosis. When the illness lingers over months or years, families can suffer isolation, lonely struggle, and exhaustion. For many parents this is the 'dying period'. The working party found the concept of a 'dying period' useful, although they realized that the dying period and final illness cannot always be separated (Fig. 20.3).

Support is needed from the time of diagnosis. A relationship between supporting professionals and the family needs to be built up at an early stage. It should be maintained, at varying intensities, throughout the illness, by familiar people. An influx of strangers to help at the final illness is not appropriate.

Improving communication

One of the greatest weaknesses of the present system lies in inadequate communication at every level, for instance between hospital and community doctors, statutory and voluntary services, and the health and education services (Fig. 20.4). There appears to be an inability for professional groups to step outside usual boundaries and to abandon traditional hierarchies. Thus,

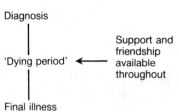

Fig. 20.3   Continuity in support and care

although concern and goodwill are present everywhere, they are not being harnessed and families are suffering needlessly. In the words of one mother,

'There was a dearth of information. You pick up information from newspapers and other parents. Then you start to fight. There is such confusion amongst the statutory organizations as to who is responsible for what, you get tossed from one to another and end up with nothing. We have had six years and in that time we have built the team up around us.'

Figure 20.4 represents the sort of supporting structure which that increasingly assertive and able family had built. It had taken six years because it had depended on reaction rather than planning. Less demanding families live in isolation and inadequacy because there is no one person with the responsibility of checking that their support structure is in place.

The working party felt that each family should have such a key person, chosen by them from amongst the professionals involved. This person should make him/herself well-known to all members of the caring team and should have the following role:

- to direct parents to sources of information and help;

- to serve as a liaison person, offering contacts and ensuring that the services actually materialize;

- to be sympathetic, but not necessarily to have a counselling role; to direct parents to appropriate counselling, as required;

- to act as an enabler, endeavouring to make all relevant parts of the structure accessible and preventing recourse to confrontation.

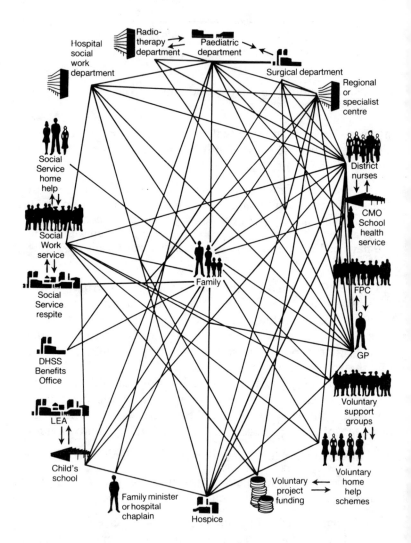

Fig. 20.4 Organizations and agencies likely to be involved with one family. (Figs 20.1–20.4 reproduced by kind permission of the National Association of Health Authorities (NAHA) from their guidelines. *Care of Dying Children and Their Families.*)

Strengthening community care

The working party felt that a primary duty of health authorities was to develop and implement a policy of supportive community care, with the aim of keeping families together and enabling parents to continue looking after their children at home. In the report, ways of managing the service, the role of domiciliary nurses, respite and short relief care, co-operation with trusts, and other fund-raising bodies are all discussed.

Domiciliary nurses
Domiciliary nurses would form the core of the service. Allocated to each family at the time of diagnosis, their job would have the following elements:

- listening and enabling;
- advising on nursing matters and assisting when needed;
- sharing information and ensuring that the parents understand and are coping;
- helping the parents to solve family problems arising from the illness;
- liaising with the key person or acting as an access point to other services;
- continuing support after the child's death.

The degree of support will depend on the type of illness, stage of disease, and strength of the family; but the nurses should continue occasional visits when all is going well and they should allocate enough time to sit down and talk. For those families who have little contact with the hospital, arrangements should be made to provide day and night cover by telephone.

Respite care
Respite care was seen as part of the community support service. The need for respite care, as a way of keeping families together and enabling parents to continue caring for their children at home, should not be underestimated and parents should be introduced to the idea at an early stage in their child's illness. Many families make their own arrangements or use excellent

schemes arranged by the Social Services. However, for children with a progressive degenerative condition, a problem point is likely to arise when they become severely ill or incapacitated so that these systems cannot cope with the degree of medical or nursing skills needed. For these children it is the duty of health authorities rather than Social Services to provide respite care. The working party looked at existing provisions and found that much of it did not match up to the principles laid out earlier in the report.

The philosophy of children's hospices for the provision of on-going care and respite provides a model that health authorities could use to develop an integrated community support service.

Voluntary organizations

In this field the importance of voluntary organizations, voluntary money, and trusts is paramount. They have been and still are, providing the major action and development. These organizations have realized the needs, seen gaps in provision and have acted to fill them.

The provide a service at several levels, such as:

- emotional support for individual families;
- collection and dissemination of information;
- education and training of staff;
- provision of short-term relief and help in the home;
- respite care;
- financial support of health service and social service personnel;
- fund-raising power.

They are already part of the caring system. They often provide a combination of skills that are hard to find in official agencies. Good relationships need to be developed so that their expertise and services are used to the full. No plans for a comprehensive service are possible without co-operation between the voluntary and statutory services.

An enormous amount remains to be done if there is to be a proper recognition of the needs of children with a limited life expectation and their families. In preparing this report, the working party listened to the views and ideas of many families in addition to the suggestions from professional groups. We hope that all health authorities, regional and district, will find our recommendations useful in the planning of a strategy for this neglected group of children.

References

DHSS (1987). *HC(87)4, Health Service development, terminal care.* DHSS, London.

Dominica, F. (1987). The role of the hospice for the dying child. *British Journal of Hospital Medicine*, **38**, 334–43.

Eardley, A. for the working party on the 'Care of the Dying'. (1987). *Care of the dying: a guide for health authorities.* King Edward's Hospital Fund for London and National Association of Health Authorities, Birmingham.

Salvage, J. (1986). *Hospices for children: a need in a sick society? Proceedings of a conference organized jointly by King's Fund Centre and Helen House Hospice for Children held at the King's Fund Centre on 2 December 1986.* KF No 86/181, King's Fund, London.

Thornes, R. for the working part on the 'Care of dying children and their families'. (1988). *Care of dying children and their families.* National Association of Health Authorities, Birmingham.

The way forward

FRANCES DOMINICA

There are key words which run through the different contributions to this book—loneliness, isolation, communication, liaison, expertise, support. There is a child and a family at the centre of the stage enacting a tragedy in which you are involved, expressing feelings of loneliness and isolation, of not being heard, of being made to feel at best fussy, at worst neurotic. I recently attended a Families' Conference, where there were sixty families whose child or children all suffered with spinal muscular atrophy; the unspoken agenda was anger and frustration. Anger, that they had to fight so hard all along the way for their child. Frustration, that no one wants to know any more—or so it feels—and that they have been left high and dry with the pain and the struggle of the years that their child has been dying. And it is the health-care professionals more often than not who are the focus of that anger. I know of a family where over twenty people (and the counting stopped there) are involved in provision of support through statutory or voluntary agencies, and yet at the centre of it all, the family feels unsupported, confused, and lonely. What do they need? One person who listens and hears, who recognizes *their expertise* with their own child and stays alongside and supports them in their role. That one person may have any or many training, skills, and experience, but needs to be a fully paid-up, card-carrying member of the human race, not a busy, heavily masked professional.

It is important that this book contains contributions from parents who have a child suffering from a life-threatening illness; and parents whose child or children have now died; all have generously described their personal experiences. They can write with authority of experience and say, 'I know something of how you feel'. The rest of us *cannot*. We, the professionals, are the pupils; the child and the family are the teachers: it is a master class.

In taking a professional stance and commitment to caring, there is ever the risk of self-satisfaction and complacency. It may be salutary to remember that a world exists beyond our small island, a world where more than five million children under the age of five die every year of chronic diseases other than those which we have been considering; dying of malnutrition, or diarrhoea and dehydration; in a world torn asunder by war and hatred in which arena our contention that the death of a child is an outrage seems strangely misplaced. I have a daily reminder at home. At the age of 45, I finally earned the title that I held for some years, Mother! In June 1988, I was speaking at a Clergy Conference in West Africa. I was invited to visit a hospital and there I met many children and their relatives. One child was different—he was alone. I assumed he was a few weeks old, judging by his size, (he weighed 8 lbs) but was told that he was about 10 months old. He had been abandoned at the age of 4 months and, in hospital, was failing to thrive and fading from life. Impulsively, I broke all my own rules and ignored generally held counsel. I offered to have him for a few months, to build him up: then from a position of physical strength his future could be planned. A year after our initial meeting, he weighs about 25 lbs and is radiant with health and enthusiasm for life. Children in the West African hospital are dying daily, of gross anaemia, infection, severe malnutrition, and heart failure. It looks as if this one little boy now living in Oxford will not share the daily fate of tens of thousands of his fellows.

What then, is the way forward? It is to listen to the children and their families—the experts—and to be aware of our own feelings about dying and death, our own feelings about life and living.

As professionals we must be concerned with the particular child and the particular family, who look to us for help *now*. We must give them 100 per cent of our attention. They must be able to feel and believe that for us, momentarily at least, nobody else in the world exists.

Appendix: *Glossary of organizations and services*

Entries in this glossary are either direct references from the preceding chapters or have been included because of their shared purpose or interest. Therefore, whilst it is not intended as an exhaustive directory—simply as a useful collection of organizations not previously grouped together in this way—should any reader feel that a relevant body has been overlooked, **ACT** (see below) would be grateful to have details.

ACT

(**A** ction for the care of families whose **C** hildren have life-threatening and **T** erminal conditions)—a national survey of those existing support services (medical, social, educational, and befriending) which are available to families whose children have life-threatening conditions. The survey includes educational and personal support services for professional carers.

The data, collected from statutory and voluntary/trust funded organizations and parent self-help groups, forms the basis of a National Resource and Information Centre for use by families and professionals alike. In addition to answering enquiries, the Centre is a forum for networking information, aiming to widen lines of communication within and between statutory and voluntary services and the families themselves. ACT hosts a number of multi-disciplinary meetings—to include parents—in which ideas and initiatives are discussed which will complement or develop existing services but avoid wasteful duplication of effort and resources.

ACT has a National Advisory Committee comprising representatives of major national organizations and the statutory services. In looking to the committee to support its projects and consider its findings, ACT seeks to promote a nationally co-ordinated approach to the care of these children and their families.

ACT is principally funded by Nuffield Provincial Hospitals Trust and the Beatrice Laing Trust. Additional grants are from the Hayward Foundation, Smith's Charity, Burmah Oil plc, the Hedley Foundation, the Tom and Miriam Stoppard Trust, the Rupert Foundation, and REACT.

Although at an early stage of collecting and co-ordinating information, ACT endeavours to answer all enquiries and consequently, welcomes details from any group or individual about their current or potential projects. ACT, Institute of Child Health, Royal Hospital for Sick Children, Bristol BS2 8BJ. Tel: 0272 221556.

Acorns Children's Hospice

Acorns' Care Team provides a community based service to support the families of children with life-threatening conditions. An integral part of this service is respite care offered at Acorns, a house which has ten beds for youngsters (up to the age of 19 years) who can stay for up to a fortnight at a time, with or without their family. Acorns, 103 Oak Tree Lane, Selly Oak, Birmingham B29 6HZ. Tel: 021 414 1741.

Alder Centre

For *all* those affected by the death of a child—parents, brothers and sisters, grandparents, friends, and professional carers in both hospital and community. Together with individual counselling from its full-time staff, the Centre provides a meeting place for parent self-help groups, and, if wished, an opportunity for newly bereaved families to meet outside the Centre with others less recently bereaved who are ready to offer support. Parents run an evening telephone helpline. Celia Hindmarch, Manager/ Counsellor, Alder Centre, Royal Liverpool Children's Hospital, Alder Hey, Eaton Road, Liverpool L12 2AP. Tel: 051 228 4811 ext 2391.

BACUP (British Association of Cancer United Patients)

A national information service whose team of specially trained nurses provides information, practical advice, and emotional support to cancer patients, their families, and friends, by telephone or letter. A comprehensive library and computer resource directory has been developed to provide essential information for BACUP's nurses and for outside medical and health professionals. A wide range of easy-to-understand publications on most types of cancer, specially written for patients by professionals, is available. A quarterly journal *BACUP News* keeps individuals up to-date with developments within BACUP and the cancer field in general. 121/123 Charterhouse Street, London EC1M 6AA. Tel: 071 608 1661. (Freeline for callers outside London 0800 181199.)

Barnardos Cystic Fibrosis Project (West Midlands)
The project provides a welfare and social work service for those with cystic fibrosis and their families, throughout the West Midlands. Anyone may refer although most referrals are from consultants at the Birmingham Children's Hospital. Project Leader—Tina Neale, Barnardos CF Project, 90 Broad Street, Birmingham B15 1AU. Tel: 021 633 3522.

Barnardos Cystic Fibrosis Project (East Midlands)
This project has recently been developed to provide the East Midlands with a similar service to the West Midlands project. Project Leader—Sheila Hepple, Barnardos East Midlands CF Project, Rufford, Nottingham City Hospital, Notts NG5 1PB. Tel: 0602 691177.

Barnardos St James Project (Leeds)
A counselling and support service for those children with specific life-threatening illness and for their families. These conditions currently include cystic fibrosis, renal, liver and chronic bowel disease. Project Leader—Heather Collington, St James's University Hospital, Leeds LS9 7TF. Tel: 0532 433144. ext 5408 or 0532 582115

Barnardos Orchard Project (North East)
Provides help and support to families and young people facing loss or bereavement. In addition a social worker provides a service for families and children with cystic fibrosis. Project Leader—Wyn Ross, Orchard House, Fenwick Terrace, Jesmond, Newcastle-upon-Tyne, NE2 2JQ. Tel: 091 281 5024.

Bristol and South West Children's Heart Circle
Providing help and support to families of children with heart problems, especially when admitted to hospital for investigation or cardiac surgery. Through monies raised, the Circle has been instrumental in providing social and medical amenities which include twenty-four bed sitting rooms for parents staying in hospital with their children, and funding for the post of Paediatric Cardiac Counsellor. A growing number of parent self-help groups throughout the region offer friendship and support. Chair: Mrs Jean Pratten, 19 Coldharbour Road, Redland, Bristol BS6 7JT. Tel: 0272 734373. Details of similar groups throughout the UK may be obtained from Mrs Fiona Benson, Secretary, Heart Care, 6 Waterloo Road, Bedford MK40 3PQ. Tel: 0234 50030.

British Association for Counselling
A national organization which promotes a wider understanding and awareness of counselling; increases the availability of counselling by trained and supervised counsellors; maintains and raises standards of counselling training and practice; provides support for counsellors; offers information and advice concerning both counselling and counsellors; represents counselling at a national level. 37a Sheep Street, Rugby, Warks CV21 3BX. Tel: 0788 78328.

CALL
(Childhood Cancer and Leukaemia Link)—a national organisation providing family to family support for those affected by childhood cancer or leukaemia. Membership is free to affected families and informal groups operate locally with a quarterly newsletter providing a forum for discussion and exchange of information on a national basis. Families can be linked with others in a similar area or situation. CALL is run by families who have experienced childhood cancer but also links with professionals working in this field. Mrs Gill Denne, 20 Haywood, Bracknell, Berks RG12 4WG. 0344 423635

Cambridge Children's Hospice for the Eastern Region
A respite care centre offering help and support to chronically or terminally sick young people (from 0 to 18) from Norfolk, Suffolk, Lincolnshire, Bedfordshire, Hertfordshire, Essex, and

Cambridgeshire. Children and adolescents with a wide variety of physical and mental conditions stay for short periods to allow their families 'breathing space', although, if they wish, the families are most welcome to stay. Cambridge Children's Hospice, Milton, Cambridge CB4 4AB. Tel: 0223 860306.

CancerLink

A national organization providing emotional support and information in response to enquiries on all aspects of cancer from people with cancer, their families and friends, and from professionals working with them. Acting as a resource to over 300 cancer support and self-help groups throughout Britain, and helping those setting up new ones. 17 Britannia Street, London WC1X 9JN. Tel: 071 833 2451.

Cancer Relief Macmillan Fund

A national charity helping patients with cancer and their families; financially assisting patients with grants, providing monies for hospice and home care. Initially funding Macmillan Nurses (see Macmillan Nurse) and promoting a better understanding among professionals of needs of patients and carers. Cancer Relief Macmillan Fund, Anchor House, 15/19 Britten Street, London SW3 3TZ. Tel: 071 351 7811.

CLIC (Cancer and Leukaemia in Childhood Trust)

Initially focused on the south-west region, CLIC's concern is to ensure that young patients and their families are helped from the day of diagnosis, throughout the long periods of treatment, and afterwards. Its threefold aims are Treatment, Welfare and Research through home-from-home accommodation near the paediatric oncology units, crisis-break holiday flats, teams of domiciliary care nurses and a number of research posts and projects. 1990 will see the launch of CLIC UK to offer advice and tangible assistance to any group in the country wanting to follow CLIC's example as a model of care. CLIC House, 11/12 Freemantle Square, Bristol BS6 5TL. Tel: 0272 248844.

The Compassionate Friends

An international organization of bereaved parents offering friendship and understanding to other bereaved parents through individual or group support. Literature includes a postal library, a range of leaflets and a quarterly newsletter. 6 Denmark Street, Bristol BS1 5DQ. Tel: 0272 292778.

Contact a Family
A national charity linking parents of children with special needs through local, multi-handicap groups and national organizations providing support to children suffering from specific or rare conditions. An information, advice, and support service is available for professionals as well as individual families and groups. Regular training days are provided to assist in setting up self-help groups. 16 Strutton Ground, London SW1P 2HP. Tel: 071 222 2695.

Cruse—Bereavement Care
A nationwide organization offering counselling to all bereaved people. Cruse House, 126 Sheen Road, Richmond, Surrey TW9 1UR. Tel: 081 940 4818.

Cystic Fibrosis Research Trust
A national charity financing research and improved methods of treatment; helping to promote earlier diagnosis; supporting and advising parents caring for CF children; increasing public awareness and understanding of the disease. Alexandra House, 5 Blyth Road, Bromley, Kent. Tel: 081 464 7211.

Foundation for the Study of Infant Deaths
The major funder of research into cot death in the UK, the Foundation also acts as an information centre for parents and professionals and for the exchange of knowledge internationally. Personal support to bereaved families is offered from both the London office and local parent groups. 35 Belgrave Square, London SW1X 8QB. Tel: 071 235 0965. Ansaphone 071 235 1721.

Friends of Shanti Nilaya UK
A link with Shanti Nilaya USA founded by Elizabeth Kubler-Ross, internationally known for her pioneer work in the field of death and dying. The Friends UK hosts a series of lectures and workshops with Dr Ross and her associates, including an annual five-day residential workshops, 'Life, Death and Transi-

tion' attended by a variety of people of varying cultural, religious, and professional backgrounds, as well as by terminally ill patients and parents of dying children. Tapes and books by Dr Ross are available. 10 Archery Fields House, Wharton Street, London WC1X 9PN. Tel: 071 837 9796.

Genetic Interest Group (GIG)
A national organization to improve services for people with genetic disorders. A telephone helpline links individuals with regional genetic centres and self-help groups; seminars are held on related issues; a newsletter is published quarterly. GIG, 16 Strutton Ground, London SW1P 2HP. Tel: 071 222 2695.

Helen House
Respite care is provided for children with progressive life-threatening disease. Children coming for a first visit are normally between the ages of 0 and 16 years, but continuing support is offered beyond this age. Family members are welcome to stay with the child, sharing in their care if they so wish. The environment is homely and informal with room for a maximum of eight sick children at any one time.

Help and support are offered during the terminal phase of illness, whether at home or at Helen House, and through the period of bereavement following the child's death. The pattern and type of care offered throughout is flexible according to the needs of the individual child and family. Helen House, 37 Leopold Street, Oxford OX4 1QT. Tel: 0865 728251.

Leukaemia Care Society
A national organization to promote the welfare of people suffering from leukaemia and/or allied blood disorders, together with their families. The Society provides support through information, financial help, and holidays. PO Box 82, Exeter, Devon EX2 5DP.

Leukaemia Research Fund
A national charity supporting research work for leukaemia and related blood diseases at its own specialist centres and at many hospitals and universities throughout Britain. Publications include a yearbook, newsletters, and patient information book-

lets. 43 Great Ormond Street, London WC1N 3JJ. Tel: 071 405 0101.

Lisa Sainsbury Foundation
An education and information service for health professionals caring for the terminally ill. Workshops are offered, tailor-made to meet the precise needs of a small group; popular topics are; talking with patients and relatives, communication, counselling skills, bereavement and loss, pain relief and symptom control. The Foundation publishes a series of paperback books, a list is available on request. 8–10 Crown Hill, Croydon, Surrey CR0 1RY. Tel: 081 686 8808.

Macmillan Education Centre
A hospice-based, multi-disciplinary educational resource for professionals caring for the dying patient of any age, and the family. The Centre offers support, advice, and education to those professionals already involved in caring for dying patients, and also to those interested in setting up new services in the UK or abroad. Dorothy House Foundation, 164 Bloomfield Road, Bath BA2 2AT. Tel: 0225 44545.

Macmillan Nurse
A clinical specialist, based either in hospital or the community, who is a resource to professionals caring for patients with cancer, and to their families. Initially funded by Cancer Relief Macmillan Fund (see Cancer Relief). Macmillan Nurses, Cancer Relief Macmillan Fund, Anchor House, 15/19 Britten Street, London SW3 3TZ. Tel: 071 351 7811.

Malcolm Sargent Social Worker
A specialist social worker funded by the Malcolm Sargent Cancer Fund for Children, working exclusively with children suffering from any form of cancer, and their families, as team members of most of the major regional paediatric oncology units throughout the UK. They are involved from the time of diagnosis until the time of discharge or of the death of a child—and beyond, to provide bereavement support and counselling. Malcolm Sargent Cancer Fund for Children, 14 Abingdon Road, London NW8 6AF. Tel: 071 937 4548.

Marie Curie Nursing Service
In conjunction with health authorities around the country and with some hospices and home care teams, Marie Curie Nurses assist in caring for terminally ill cancer patients who are being looked after at home. They will stay with them for long periods of time either during the day, or more commonly, at night, undertaking any practical care required. Through their support relatives are able to obtain rest or take a break outside their home. There is no charge to patients. Marie Curie Memorial Foundation, 28 Belgrave Square, London SW1X 8QG. Tel: 071 235 3325.

Martin House
Providing respite care for children suffering from life-threatening disorders, and for the whole family, in a home-from-home, friendly atmosphere where special needs can be looked after by qualified people. Normally, children will be accepted between the ages of 0 to 16 on first referral. The Care Team also provides round the clock help and support at home. The continuing work of support for bereaved families is willingly accepted. Martin House, Grove Road, Clifford, W.Yorks, LS23 6TX. Tel: 0937 845045.

Meditec
A specialized bookselling service covering all aspects of care of the dying, bereavement, grief, counselling, and oncology nursing for patients, families, and professionals. A regularly updated stock list is available free on request. World-wide mail order service provided with UK orders sent post free. Meditec, York House, 26 Bourne Road, Colsterworth, Lincs NG33 5JE. Tel: 0476 860281.

Miscarriage Association
To provide information and support for women and their families during and after miscarriage. PO Box 24, Ossett, W. Yorks WF5 9XG.

MPS Society for Mucopolysaccharide Diseases
A national society which aims to act as a support group for parents of children suffering from these disorders, to bring about

a greater public awareness of MPS diseases and to raise funds for research into cause and treatment. Activities include an advice and information centre for parents and professionals, an annual conference for parents, and holidays for groups and families. MPS, 30 Westwood Drive, Little Chalfont, Bucks HP6 6RJ. Tel: 0494 762789

Multiple Births Foundation Bereavement Clinic
As a facility of the Foundation, this Clinic is held every three months at which parents may have individual appointments with the paediatrician and/or meet with other parents. Similarly, a Special Needs Clinic, is available for parents of twins, one or both of whom have difficulties, such as chronic illness or handicap. Queen Charlotte's and Chelsea Hospital, Goldhawk Road, London W6 0XG. Tel: 081 748 4666 ext. 5201.

Muscular Dystrophy Group of Great Britain and Northern Ireland
A national organization funding research into the cause and cure of muscular dystrophy and allied diseases and providing resources, through conferences, seminars and literature, for the education of professionals and families in the management of neuromuscular diseases. Assistance is afforded to those with muscular dystrophy and their carers, linking them with existing statutory or voluntary services or through support from the Group's own Domiciliary Care Teams. Publications include a monthly newsletter *In Focus* and a bi-annual review, *The Search*. M D Group, Nattrass House, 35 Macaulay Road, London SW4 0QP. Tel: 071 720 8055.

National Association of Bereavement Services
Set up in 1988 to co-ordinate the many unsupported and often isolated bereavement services which exist throughout the country. The Association's aims include: compiling a national register of existing services, highlighting gaps and encouraging the provision of new services, providing a forum for members, promoting high standards of training (to include training courses) and practice for local bereavement services. 68 Charlton Street, London NW1 1JR. Tel: Ansaphone 071 388 2153.

National Association of Family Based Respite Care

Formed in response to a need for a central body to represent the views of parents, carers, users, and professionals, the Association keeps a comprehensive register of family based respite care schemes for people with a severe disability. It provides information to help and encourage development of new schemes and intends setting up a code of practice as a guideline. Christopher Orlik, National Research and Development Officer, The Nora Fry Research Centre, 32 Tyndall's Park Road, Bristol BS8 1PY. Tel: 0272 238566.

National Association for the Welfare of Children in Hospital (NAWCH)

Supporting sick children and their families and working to ensure that health services are planned for them, NAWCH passes on guidance about the needs and care of children to parents, professionals, and policy makers through their national information service, publications, and local branches. NAWCH, Argyle House, 29–31 Euston Road, London NW1 2SD. Tel: 071 833 2041.

National Children's Bureau (NCB)

Established to promote the benefit and welfare of children and young people; to organize co-operation in achieving this objective by bringing together representatives of statutory authorities and voluntary organizations; to promote and carry out research, surveys, investigations; to provide an information service; to publish and disseminate knowledge and experience; to arrange and assist in promoting conferences, seminars, meetings, and discussions.

The National Children's Bureau assists in the development of policy and good practice in services for children, contributes to national debate and identifies problems and issues which affect the welfare, education, and health of children and young people. It is multi-disciplinary in its approach, pursues an equal opportunities policy and is concerned to enable the voices of children and young people to be heard and taken into account in all its work and activities. 8 Wakley Street, London EC1V 7QE. Tel: 071 278 9441.

National Self Help Support Centre
A network linking together people who work with self-help groups locally. It acts primarily as an informal support system for the workers. It meets once a year to share information, strengthen links and break the isolation of local workers. National Self Help Support Centre, NCVO, 26 Bedford Square, London WC1B 3HU. Tel: 071 636 4066.

Perinatal Bereavement Research Unit
A National Health Service unit in the Tavistock Clinic, concerned with training and clinical research in stillbirth and similar loss around pregnancy. Six to eight week courses on perinatal bereavement (one day per week) are held each academic term for health care professionals. Perinatal Bereavement Research Unit, Tavistock Clinic, Tavistock Centre, 120 Belsize Lane, London NW3 5BA. Tel: 071 435 7111.

The Rainbow Centre for Children with Cancer and Life-threatening illness
A national centre offering therapeutic support and counselling to these sick children, their parents, and brothers and sisters, as well as to the family as a whole. Advice is available on complementary therapies, such as relaxation/visualization, diet, and vitamin and mineral supplements, which may be used alongside conventional medical treatment.

The Rainbow Centre also offers counselling and support for families or individuals who are affected by the death of a child, including babies who have died as a result of miscarriage, termination, or stillbirth, or where children are facing the loss of a parent or sibling. PO Box 604, Bristol BS99 1SW. Tel: 0272 730752 (Therapies); 0272 736228 (Administration).

Rainbow Trust
A crisis organization whose team of domiciliary workers offers practical and emotional support on a round-the-clock basis, to a family whose child has a life-threatening disease. It also provides a temporary respite holiday home—Rainbow House—for a maximum of three children and their families. Rainbow Trust, Wyvern House, 1 Church Road, Leatherhead, Surrey KT23 3PD. Tel: 0372 59055/59033.

REACT (Research, Education and Aid for Children with Terminal disease)

Registered as a national charity in 1990 to assist children with life-threatening illness and their families when faced with un-expected, financial need. REACT attempts to respond immedi-ately to applications from parents (submitted through their doctor or other professional workers). As resources become available, REACT will also increasingly fund research into the treatment of these children, and the problems faced by those involved with their care. 73 Whitehall Park Road, London W4 3NB. Tel: 081 995 8188.

RTMDC (Research Trust for Metabolic Diseases in Children)

To raise funds to further medical research into the one thousand different incurable metabolic diseases; to offer parents support through three yearly newsletters, an annual conference and a parent counsellor (Mrs Lesley Greene) on 24-hour call to advise, inform, and support parents of the newly diagnosed, and the bereaved. Contact with other families can be made and visits arranged. RTMDC, 53 Beam Street, Nantwich, Cheshire CW5 5NF. Tel: 0270 629782; 24-hour counselling line 0270 626834.

Rupert Foundation

To provide assistance to children and their families where a child is suffering from a life-threatening disease, by sponsoring teams of specially trained nurses in National Health Service hos-pital units where approximately eight such children are receiving treatment. The Rupert Foundation, Upton, East Knoyle, Salis-bury, Wiltshire SP3 6BW.

SANDS (Stillbirth and Neonatal Death Society)

A national organization offering support through self-help groups and befriending to bereaved parents who have suffered a stillbirth or neonatal death. Aims include increasing awareness among health care professionals and the general public of the needs and feelings of bereaved parents. SANDS, 28 Portland Place, London W1N 4DE. Tel: 071 436 5881.

SATFA (Support After Termination For Abnormality)
A national self-help charity run by women and couples who have experienced a termination because an abnormality was diagnosed in their baby. Aims are to establish a national network of support groups as well as working with health professionals to encourage a greater understanding of needs of parents at this time. 29–30 Soho Square, London W1V 6JB. Tel: 071 439 6124.

Sick Children's Trust
Formed in 1981 by two hospital consultants to provide home-from-home accommodation for the families of children who are seriously ill in hospital. Currently, the Trust has three such centres open in London with plans for another to be attached to Moorfields Eye Hospital. Further homes-from-home are scheduled to open in Kent, and the north of England. 10 Guilford Street, London WC1N 1DT. Tel: 071 404 3329.

Tadworth Court Trust
A registered charity for children offering special non-residential education, respite care, rehabilitation, and treatment for those who have sustained head injuries or are profoundly multi-handicapped or are chronically sick—especially with the management of cystic fibrosis or degenerative conditions. A full range of therapy disciplines is available to each child. During the summer months, holiday activities are provided for those children who are receiving respite care. A family counselling service offers parents the opportunity to share their problems in privacy. Hospice terminal care and family support is also available. Children maybe referred from any area by parents or professionals in the first instance. Sponsorship/funding is usually obtained from statutory authorities. Gill Meyer, Liaison Sister, Tadworth Court Trust, Tadworth, Surrey KT20 5RU. Tel: 0737 3571.

TAMBA (The Twins and Multiple Births Association)
A national support organization for families with twins or other multiple births, also providing information to professionals. There are many sub-groups including the Bereavement Support Group for families who have suffered the loss of one or more of a set of multiples, from early miscarriage to the death of

an older child or children. General enquiries should be sent with a stamped addressed envelope to Mrs Mary Lowe, 51 Thicknall Drive, Pedmore, Stourbridge, W. Midlands DY9 0YH. Bereavement Chairman, Mrs Shelley Eisen, 56 Chase Court Gardens, Enfield, Middlesex EN2 8DJ.

Index